KHOSROW M. HADIPOUR

SAND CONTROL AND GRAVEL PACKING TECHNIQUES

It Never Rains in the Oil Field!

To order additional copies of this book, contact:
Xlibris
1-888-795-4274
www.Xlibris.com
Orders@Xlibris.com

Sand-Control and Gravel-Packing Techniques

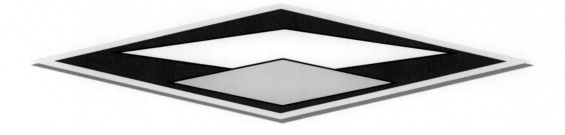

It Never Rains in the Oil Field!

Published by Khosrow M. Hadipour, professional petroleum engineer, Texas A&M University

The subject material is based on forty-one years of offshore and onshore downhole experience in drilling, completion, production, fracturing, downhole fishing, sidetrack drilling, cementing, coiled tubing operation, oil-and-gas remedial workover repairs, artificial fluid lift, gravel packing, and plug/abandonment while working for the Gulf Oil company, Chevron USA, Pennzoil Company, Devon Energy, and AmeriCo Energy Resources in Texas, Mississippi, Louisiana, New Mexico, and Venezuela.

It is not the purpose of this book to be used as final procedure and/or definite guideline. This book is prepared to act as basic reference based on my experience only.

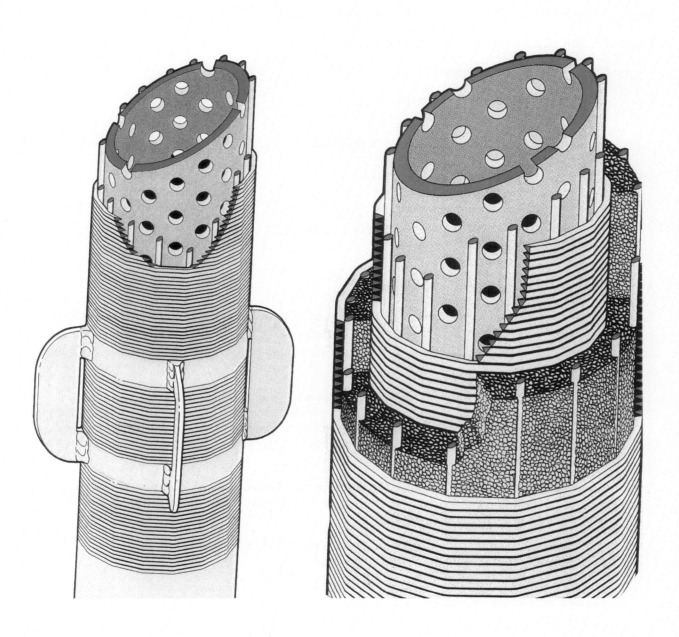

Gravel-Pack Sand-Control Technology in Oil-and-Gas Fields

Gravel Packing is an art of controlling formation sand and solids from entering into a wellbore. Formation sand and solids are one of the costly unwanted materials coming out of oil-and-gas wells that must be kept under control. The most effective method of controlling formation sand and solid is gravel packing immediately upon initial well completion.

Slight formation disturbance will result in reduction of productivity.

There are basically two sand-control completion techniques:

A. Conventional sand-control completion methods

- open-hole sand-control completion
- case-hole sand-control completion

B. Standard gravel packing sand-control methods

- open-hole gravel-packing completion techniques
- cased-hole gravel-packing completion techniques
- through-tubing gravel-packing techniques

****

A.   Conventional Sand-Control Methods

The concept of conventional sand-control techniques is done based on the coarse and larger sand-gravel that may be produced by the reservoir formation. This method of well completion has been done since the early days of oil field in the open-hole and/or the gun-perforated cased-hole sand-control methods.

1. In the conventional open-hole sand-control method, the wire-wrapped screen and liner is designed based on the reservoir coarse sand grain samples from the open-hole formation. The wire-wrapped screen and liner is run and set at the bottom in the open hole directly opposite of the productive reservoir in the open hole and extends upward into the steel casing above the open-hole section. The annulus will be isolated in the casing section using a completion packer.

   The open-hole sand-control completion technique is an excellent method of obtaining maximum allowable production out of the reservoir. The open-hole sand-control completion is applied in oil wells, saltwater disposal wells, and freshwater wells.

   The concept of conventional well completion in an open hole is subject to change because of multiple zones with shale streaks above the productive reservoir.

   The shale strata may cave in and cause failure in an open hole, causing gravel-pack failure and fishing problems. In this case, gravel packing around the screen and liner joints may be necessary to keep the shale strata from caving in around the screen and liner, causing sand-control failure.

2. In the screen-hang-off method, prepacked screen is run, hung off, or simply set at the bottom of open perforations using an isolation packer to protect the screen/liner.

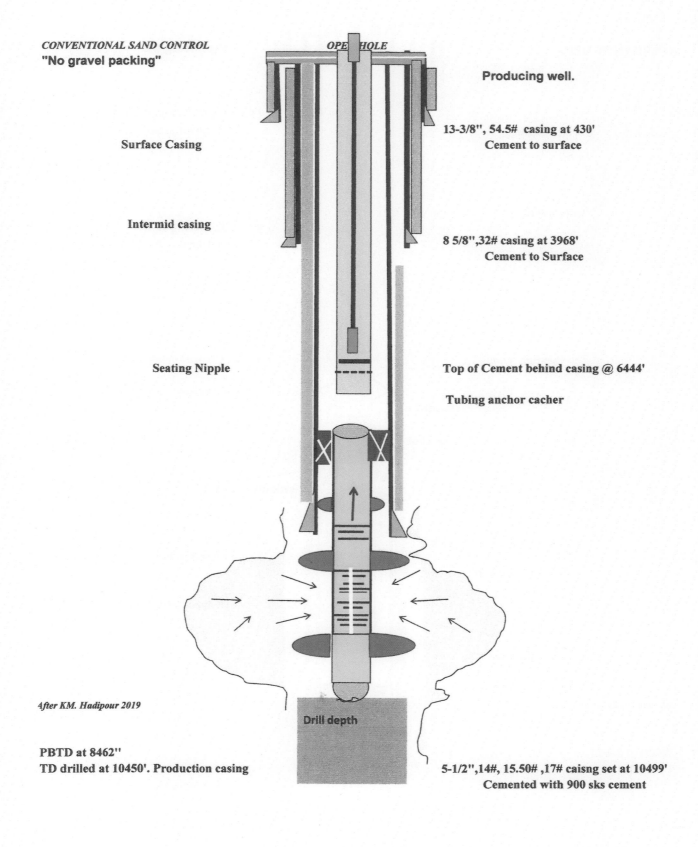

CONVENTIONAL SAND CONTROL
"No gravel packing"

OPEN HOLE

Producing well.

13-3/8", 54.5# casing at 430'
Cement to surface

Surface Casing

Intermid casing

8 5/8",32# casing at 3968'
Cement to Surface

Seating Nipple

Top of Cement behind casing @ 6444'

Tubing anchor cacher

After KM. Hadipour 2019

Drill depth

PBTD at 8462"
TD drilled at 10450'. Production casing

5-1/2",14#, 15.50# ,17# caisng set at 10499'
Cemented with 900 sks cement

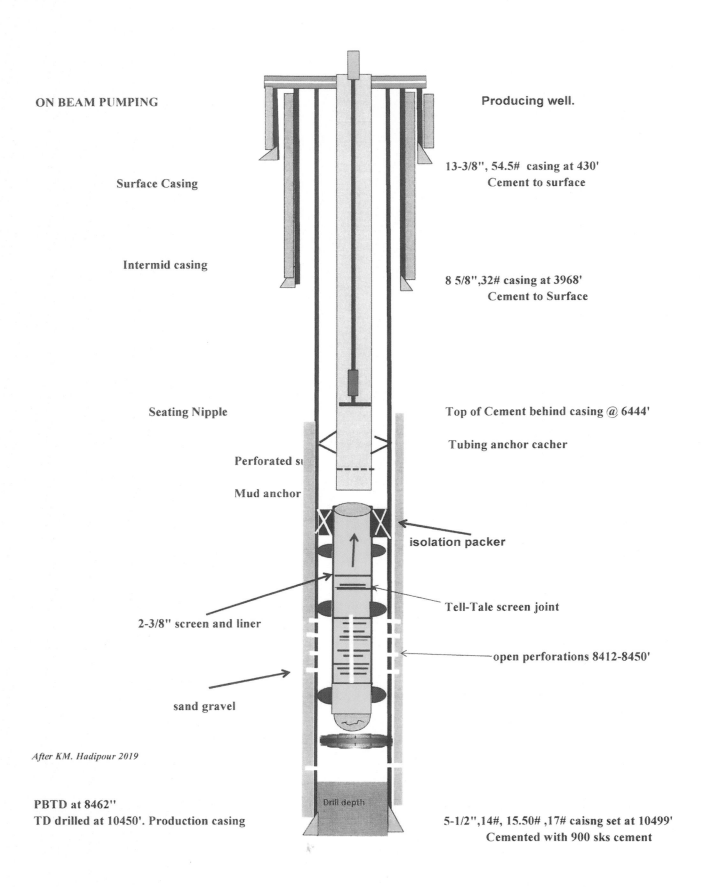

ON BEAM PUMPING

Producing well.

13-3/8", 54.5# casing at 430'
Cement to surface

Surface Casing

Intermid casing

8 5/8",32# casing at 3968'
Cement to Surface

Seating Nipple

Top of Cement behind casing @ 6444'

Tubing anchor cacher

Perforated su

Mud anchor

isolation packer

Tell-Tale screen joint

2-3/8" screen and liner

open perforations 8412-8450'

sand gravel

After KM. Hadipour 2019

PBTD at 8462"
TD drilled at 10450'. Production casing

Drill depth

5-1/2",14#, 15.50# ,17# caisng set at 10499'
Cemented with 900 sks cement

The Conventional Sand-Control and Well Completion in a Gun-Perforated Well

The concept of conventional sand-control completion in a gun-perforated wellbore is similar to the conventional open-hole sand-control method. The screen and liner is designed and run into the well based on natural coarse sand gravel that is produced from the reservoir formation without additional cost of gravel packing (a shortcut on gravel packing to save cost).

Cost cutting is an obstacle preventing you from doing the job right. (It might cost you several folds later!)

This concept of conventional completion without gravel packing was done in the early oil well sand-control methods and may be in practice in the oil fields of today.

The wellbore will be circulated clean and clear to the top of the bridge plug (foot base). A selected size of wire-wrapped screen and liner will be run and set at the bottom and across the casing's perforated holes without additional man-made gravel-packing sand. The screen and liner will be isolated with a production packer, and the well will be put on production without man-made gravel-packing sand around the screen and liner.

This method is done based on using the natural coarse formation sand that is produced by the reservoir formation to become packed off around the screen and liner without additional gravel packing. Installing the wire-wrapped screen without gravel packing in a cased-hole wellbore is impossible to justify.

The success of conventional gravel packing in a cased-hole completion may not be too favorable because of early screen-section failure (plugged-off screen). That is, the screen becomes plugged off with contaminated mud and fine solids, and it may cause a large decline in the productivity.

B.   The Standard Gravel-Packing Completion Technology

The standard gravel-packing technique is applied in a cased-hole gravel packing, as well as the open-hole gravel packing.

1. Run a selected sized screen and liner into a well and pump gravel-packing sand around the screen and liner.

2. Prepack the reservoir formation with sand-gravel first, then run and set the screen and liner in place across the open perforation and pump sand-gravel around the screen/liner. (It is carried out by a crossover method or a wash-down method.)

3. The through-tubing gravel packing (another innovating gravel-packing completion technique offered by the Gulf Oil company to save production) is normally applied in the wells with formation-sand problems, but it is too costly to use a workover rig to work on the well.

What Is Gravel Packing?

Gravel packing may be defined as "the placement of selected resieved sand-gravel behind and across the gun-perforated holes and around the sand screen to stop the migration of formation sand entering into the wellbore." The idea of gravel packing is actually derived from freshwater well sand-control techniques to the oil-and-gas wellbore gravel packing and the sand-control methods of today.

The technology of sand control and gravel packing is well-known throughout the oil industry and the underground freshwater well completions since the 1900s. Nearly all the municipal water-supply wells are gravel packed in order to sell clear and clean water to the public.

The basic idea behind oil-well gravel packing is to create a form of screen or filtration mechanism between the reservoir formation and the cased-hole open perforations to block off or hold back the reservoir's unconsolidated formation sand from entering into the wellbore.

The gravel-packing sand is derived from crushed, reprocessed, manmade silica gravel with higher permeability and porosity than most reservoir formation sands. The formation fluid will travel through the gravel-packing sand at a lower velocity and may reduce turbulence and formation disturbance.

Gravel packing is an art of filtering the reservoir fluid to control unconsolidated formation sand.

Sand filtration will take place as soon as high-porosity gravel-packing sand is forced across the reservoir formation. Gravel-packing sand will hold back the formation sand. The screen/liner will frame and support the gravel-packing sand and the formation sand in its original place.

The Elements of Surprise

Production fluid coming out of an oil-and-gas well may consist of several visible and invisible elements:

- Crude oil (rock oil): A viscous organic compound that appears in dark, black, green, brown, and reddish color and texture. It smells like perfume to me as an oilman (believe it or not!).

- Hydrocarbon natural gas: Naturally occurring element consisting primarily of methane (CH_4) and other mixed gases, such as ethane, propane, butane, nitrogen, hydrogen sulfide (H_2S), helium, argon, carbon dioxide (CO_2), and other elements. Natural gas is colorless at low concentration. Natural gas is a valuable fossil fuel product.

- Saltwater: Unwanted dense, low viscous product with various densities.

- Scale elements: Calcium carbonate, calcium sulfate, barium sulfate, iron sulfate, and norms (naturally occurring toxic, radioactive materials that are transported to the deserts of Nevada or in your backyard).

- Drilling mud and fluid loss material: Unwanted, unstable manmade material that will break down and coagulate when mixed with saltwater (oil-based mud or water-based mud).

- Saturated salt: A hard, heavy, clear crystal or powder-like element.

- Formation sand: Unwanted loose material consisting of round- or angular-shaped quartz, feldspar, silica, and cement look-alike (like beach sand with a different physical and chemical structure and color).

- Silts and iron rust: Light, floating substance.

- Clay substance: Red-bed, blue, and greenish in color, and may appear like gumbo putty.

- Shale formations: Unstable formation that's gray, brown, black (with different colors and texture).

- Other minor elements seen coming out of a well: Coal, lignite, wood, pyrite, metallic elements, and petrified marine remains. (The elements suggest that the earth has been rolled over and tumbled once upon a time and may happen again!)

All the subject elements are absolutely unique in design and derived from dense, compacted, pressurized zones with medium to very hot temperature.

Each and every one of the above elements has special physical and chemical characteristics. The produced elements are important parts that hold the formation's structure in place. Each and every one of the above elements will expand and contract with temperature and pressure (plasticity and elasticity, expansion and contraction).

When a well is flowing naturally or artificially lifted up from a borehole, the above-mentioned elements will continue to flow in special suspended patterns in the oil, gas, and water out of the reservoir rock and will be forced up through a tubing string and onto production facilities. As the produced material is lifted up the wellbore, they may change in physical structure because of the change in temperature, pressure, and friction. Some elements may come out of liquid solution and re-form from a liquid state to a semisolid material that will stick and bond onto the wall of the tubing and casing string and production vessels in different forms (salt, paraffin, or crushed compound materials in the form of hard scale).

The harder you pull the fluid out of a reservoir; the more sand and solids will come out of a well with the fluid. (It will create large sand washouts opposite of the gun-perforated holes, which cause shifted or parted casing.)

The produced material coming out of a reservoir formation will partly or totally become separated while passing through various production vessels.

The saltwater, formation sand, mud, and other solids are unwanted by-products of the produced oil-and-gas that must be disposed properly by different techniques. (You cannot place them back to their original state, even if you wanted to.)

Based on the reservoir characteristics, some oil wells may produce a high volume of saltwater, along with the unconsolidated formation sand, while other wells produce low concentration of formation materials (it depends on overburden compaction and rock-cementation characteristics).

Practically, it will be difficult to flow or artificially lift a well fluid without unconsolidated reservoir formation in the Gulf Coast area or some of the oil fields around the world.

The unwanted microsolids in the oil will be carried out with oil products all the way to the refinery level. (It is no surprise if you find microsolids at the bottom of your gasoline tank!)

The idea behind gravel packing is to stop or slow down the flow of unwanted sand and solids using reprocessed, manmade silica gravel and mechanical steel screens as a filtering system.

The best sand prevention is to hold the solid in their original place in the reservoir, not in the wellbore or in the surface production equipment.

Reservoir rocks with high porosity and permeability will tend to produce a high volume of water, oil, and unconsolidated formation sand (loose gravel sand).

Major Important Characteristics of a Reservoir Formation to Know

Reservoir rocks contain major fluid elements, such as oil, water, or gas (material balance).

Oil and water are in liquid forms and are not compressible (referred to as Newtonian fluid).

Natural gas is a compressible fluid under pressure (LPG [liquefied petroleum gas] and LNG [liquefied natural gas] are compressed to a liquid form under a certain pressure and temperature).

Oil, gas, and water are referred to as fluid.

Fluid flows through a reservoir formation rock with good porosity and permeability (through the pores between the sand grains).

- Porosity: This is the measure of the amount of hydrocarbon oil-and-gas that is held in the rock pores. (The porosity is also referred as the pore volume per unit volume of reservoir rock.)

- Permeability: This is the measure of fluid ability to move through the sand-grain pores.

- Oil saturation: This is basically the measure of percentage of hydrocarbon in the buck volume of reservoir rock. High water saturation in a reservoir rock means a high percentage of rock pores is occupied or is contained with saltwater.

- Resistivity: This is the measure of rock resistance to block off or slow down the flow of electric current going through the rock (significant value and measured in ohm/meter).

- Bed thickness: This is the measure of reservoir sand height in feet or meters. Reservoir sand thickness may vary from one-foot thick to several hundred-feet thick (like wells in the Middle East, Venezuela, or other parts of the world).

Material balance of a reservoir fluid consists of the following:

Oil, gas, and water—all three elements will not freely mix together (immiscible fluid).
The three elements will separate according to their gravity and densities (gas, oil, and water).

The rock's porosity is responsible for a majority of the stored hydrocarbon oil stocked beneath the earth. Most of the hydrocarbon oil (rock oil) that we are using today is trapped in the rock—pores under high pressure! The rest of our oil supply may come from cracks, traps, and the salt domes in the belly of the earth (exciting!). All the oil-and-gas wells will produce certain amounts of unwanted water, sand, silt, clay, and fine crushed solids regardless of manmade mechanical filtration or chemical controlling techniques.

The formation sand and saltwater are unwanted and costly by-products of oil-and-gas that must be separated and disposed of properly by oil companies.

There are many challenging sand-control problems in oil-and-gas wells. Sand-up production perforations and production tubing string may require a costly workover rig or coiled tubing unit to remove the sand bridge and restore production.

Lowering production rate may minimize the pressure between the formation and the wellbore and reduce sand problems. This approach may result in a reduction of fluid velocity, causing sanding up and bridging effects at the bottom and across perforations. Lowering production rate means choking the fluid, the reduction of oil, and gas production.

Increasing the fluid velocity by using an artificial lifting system (or using velocity string) may reduce sand-up problems in the well but will increase a high volume of sand production to surface equipment. High volumes of sand and water production will create surface equipment damage and disposal challenges.

The best alternative to stop sand production is to gravel pack the well immediately after initial completion to prevent wellbore damages.

Some oil fields (several flowing and artificial lifting wells) may produce as much as ninety thousand barrels of produced saltwater and nearly 1 percent of sand, silt, sludge, and clay daily (based on twelve and half years of testing and observation).

(It makes me wonder when the land is going to sink!)

The 1 percent sand and sludge at the bottom of a sample jar may not seem to be a lot at the time; however, the 1 percent sand in ninety thousand barrels of produced fluid daily will add up to several ton loads of sand per day!

All produced sand and oily sludge have to pass through the production equipment at the tank battery.

What can be done with such high volume of produced saltwater and formation sand in a solution when getting into the production vessels, such as separator, heater treaters, gun barrel, free-water knockouts, and oil-stock tanks? (Get it done.)

The ninety thousand barrels of saltwater, with large volumes of sand and sludge, are unwanted and added expenses to an oil company.

- You cannot sell it!
- You cannot drink it!
- You cannot spill it!
- You cannot afford to haul it!
- You may be able to burn it and/or properly dispose it.
- You must dispose the contents properly to protect the public and the environment.

When perforating a casing string in a well, the oil, gas, saltwater, drilling mud, unconsolidated sand, and loose crushed solids will be forced into the wellbore and lifted onto production vessels, such as production separators, heater treaters, free-water knockouts, and/or gun barrels.

Production equipment in the oil field may either be a two-phase process (separate liquid from gas) or a three-phase process (separate oil, water, and gas). In a high-fluid production field, the three-phase equipment will not work properly!

In the separation process, the gas will go through the gas line for sale. Oil, water, and sand may go through production separators then a free-water knockout, and from the free-water knockout to a large gun barrel with some floating, suspended formation sand (floating sand, silts, and clay substance). The produced elements will be separated according to gravity (gas, oil, water, and solids).

Oil will separate at the gun barrel and dump into the stock tank with traces of water, suspended sand, and mud (too much oil and solids will cause bad oil to sell).

The unwanted produced saltwater will be separated, dumped out, and forced down the water line and into saltwater disposal tanks, along with some sand, mud, and traces of oil-and-gas blanket. After a period of six months, the trace of sand will appear to fill up half of one-thousand-barrel saltwater tanks that must be cleaned up. (You will need a contained entry permit to work inside a tank.)

Every few months, the saltwater tanks and other production vessels become occupied or clogged by sand and sludge and must be cleaned up to remove and dispose the formation sand and sludge as properly as possible. The repair cost of production equipment because of sand erosion is an additional everyday cost.

After several months of fluid production, the field may produce several thousand pounds of salty, oily, and dirty sand products that must be disposed of properly (like the pyramid in Egypt).

Formation sand(came from wells)

Are we actually responsible enough to dispose of our mess properly on this planet?

The term *dispose properly* is actually overstated. No one would stick his neck out to tell you what to actually do with produced sand and sludge of that magnitude! It is still an ongoing subject in some oil fields of today (offshore and onshore operations).

We do not dispose of our mess properly. We do not have the technology or desire to get there as of yet.

Some say, "it came from the earth and can go back to the earth" that is a very wrong statement. Changing the composition of a material is not the same way as that of "came from the earth." (Look at our plastic making and other nasty messes that exist in the world—dirty garbage flying across the highways and into the rivers!)

What do you actually do with all that sand? And how much is the cost to dispose the sand properly?

- You cannot dump it on the roads.
- It is costly and time-consuming to wash it on location.
- You cannot bury the sand in someone's backyard without permission.
- If you dig the ground in some of the old fields, you may see oil stains coming out of the place.

You must dispose the sand into a well at a higher cost. (Do you actually dispose the sand?) Similar costly scenarios apply to offshore operations with a higher cost than the land operations. (I hope they do not dump the sand and sludge overboard to help cut cost—all for a promotion.)

Gravel packing high-producing oil, water, and gas wells may reduce fluid production by nearly 12 percent.

Ninety thousand barrels of total fluid production with an average oil cut of 6 percent oil production revenue may reduce the oil revenue by 648 barrels of oil per day!

If you approach the landowner and your boss with an idea to gravel pack all the wells in the field with 12 percent fluid reduction, the idea may be rejected (or you may be asking for a termination).

I however, recommend to gravel pack all the wells with formation-sand problems (keep the sand where it belongs).

Do not wait too long before gravel packing a poorly consolidated reservoir formation. Do not rearrange or disturb the formation-sand grains.

The slightest change of formation-sand grains will result in loss of productivity.

A wellbore must be gravel packed immediately upon the initial production completion.

Success and/or failure of an oil or gas well can be based on several important factors:

- Initial quality and quantity of cement bonding between the casing and formation.

  There is no substitute for good-quality initial cement bonding in any oil-and-gas well.

  Unbonded cement intervals behind a casing string is a failure and will cause corrosion holes, sand problems, parted casing, and/or major remedial cementing work across oil fields.

  The oil string must be protected with cement slurry or run a longer surface casing across all the shallow wet formation sands to prevent the exposure of non-uniform casing to corrosive formation fluid.

  Running shallow surface casings to 275 feet in some wells is just a practical joke that is approved by the state government in different states.

  The surface casing should be run and cemented in place at one thousand feet minimum depth regardless of freshwater depth. The cement must be witnessed while it is circulated to surface (sample of proof).

  A casing integrity test must be conducted more often to protect the freshwater and the environment.

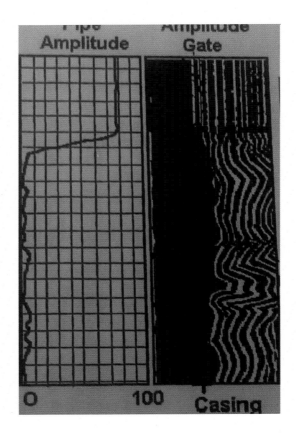

- The initial production casing string must be made dependable with high-quality steel pipe (weight and grade) to stand for longer life. (The lower grade and weight of the casing must be protected with high quality and quantity of cement slurry to resist against corrosive and hostile environment.)

  Working on some of the major oil company's wellbores across Mississippi, Louisiana, Texas, and New Mexico, I found their casing designs and well-completion techniques to be an insult to oil field engineering and the environmental protection practices.
  (Some people who are running the show in the oil fields are not qualified to work in oil fields!)

  A low-grade casing string with poor cement bonding is the first section of a casing string to fail because of corrosion and fluid erosion attack. (I worked on such wells in New Mexico, Texas, and Mississippi.)

- Look, protect your wellbore and the future freshwater by using a good quality and quantity of initial cementing around all the oil-and-gas casing strings. Do not brag about how well you plugged and abandoned your oil-and-gas wells after many years of wellbore mechanical problem and pollution. A good plug-and-abandon practice in a well after the many years of production and workovers is long overdue.

- When a long string initial cementing fails, it should be bad news for the environment and any oil- or gas-well owner!

- Perforating concept and methods—underbalanced completion and selection of large holes and shot densities are important in sand-control gravel packing to reduce deferential pressure across the selected perforations.

  Knowledge of reservoir mechanics—knowledge of mechanical characteristics of reservoir rock is very important in the completion success. Major characteristics of a reservoir may be determined from open-hole logs to learn and understand the important rock characteristics, such as porosity, permeability, water saturation, and resistivity.

  Review the fracture identification and the sonic logs to determine the rock fractures, compositions, cementation, and formation consolidation before perforating a well for gravel packing.

  High-water drive reservoirs will usually have high permeability and porosity. High water-drive wells tend to produce considerable nonconsolidated sand with water, oil, and gas production.

  Gravel packing a reservoir formation early will keep the solids in place and prevent formation breakdown, near wellbore skin damage, and casing problems.

****

Rate of Fluid Production from a Reservoir

Nearly all the high-rate flowing and/or artificial lifting oil-and-gas wells will produce sand and solids regardless of any control measures (slowing down the rate of your production).

Some formation sand will pass through the gravel-packing sand and screen with ease (slowing down the production rate will reduce sand).

Note: water-based sand-control gravel packing is not a fracturing technique!

Gravel packing is simply a placement of selected resieved silica sand through the perforation tunnels and around the screen and liner as a filter to hold back unwanted formation sand using filtered water at a low-squeezing pressure not exceeding 1,000 psi.

Artificial mechanical sand control is a method to simply slow down, reduce, minimize, block off, or hold back the produced sand and solids. (Some reservoirs may produce coarse sand grains, and some will produce very fine sand grains that can pass through any screen and liner with ease!)

Gravel-packing sand control will not be a perfect or sufficient means to totally prevent formation sand and fine solids from entering the wellbore.

Some reservoirs produce fine sands similar to baby powder (clay powder) and will pass through any screen- and gravel-pack design. Some reservoir sand may continue to penetrate through the gravel pack and will form scale blockage.

Some production reservoirs produce fluid with suspended solids that cannot be seen by the naked eye or by absolute filtration system. The suspended solids will stay in the solution and appear in the tubing and production equipment in the form of hard solid scales inside the screen and liner and the production equipment. (The color of fluid will indicate the solids in the solutions.)

Sophisticated oil-and-gas gravel-packing tools and equipment ideas were developed in 1960 and started later by Layne and Bowler, a Houston, TX. company, and the Howard Smith Company of Texas. In 1980, Howard Smith demonstrated a simple and practical gravel-packing unit and the application of thin-water gravel packing in oil-and-gas wells that is still in use today.

The purpose of gravel packing in oil-and-gas is to prevent unwanted formation sand and drilling mud from blocking off the perforation tunnels or sanding up the wellbore, prevent skin damage, and avoid casing collapse.

Unconsolidated formation sand and saltwater is a major aggravating problem in oil wells. Sand production is an expense of cleaning the wellbores and production facilities several times a year. Most Gulf Coast wells require gravel packing to avoid repeated wellbore cleaning operations.

****

The Gravel-Packing Techniques

The hydraulic method of displacing resieved silica gravel between the formation and the screen and liner is an ideal method to stop the migration of unconsolidated formation sand from entering the wellbore.

Gravel-packing sand basically comes from the crushed and reprocessed manmade silica sand.

The application of gravel-packing sand control includes the following:

- oil wells
- gas wells
- underground freshwater wells
- water-injection wells
- saltwater disposal wells
- steam flooding
- fluid filtration for any other applications

The cause of sand problems in oil-and-gas wells may be the result of the following:

1. Unconsolidated reservoir formation—The unconsolidated formation appears in various sizes and texture of sand grains with little or no bonding matrix material (broken cementation). The Gulf Coast and offshore wells produce a considerable amount of unconsolidated sand, shale, clay, and silts. Some reservoir formation produces very fine sand similar to baby powder.

2. Here are the major causes of unstable reservoir formations:

   - formation reservoir washouts during drilling operations (matrix breakdown)
   - collapse formation and sand-support bonding matrix
   - large washout and cavities behind casing and perforations while drilling
   - lack of initial cement bonding behind casing string (between formation and casing)
   - completion and perforating technology
   - stimulations and high-rate fracturing out of zone in the shallow reservoirs
   - underground blowout
   - corrosion holes in casing string across wet sand with poor cement bonding
   - matrix breakdown because of surge and swabbing techniques
   - high production rates (gas lifting, submersible, and hydraulic lifts)

3. High production flow rates in oil-and-gas reservoirs—high production rates will affect consolidated or unconsolidated reservoir sand. Maximizing fluid to increase higher oil-and-gas production will tend to cone water and produce high concentration of sand and solids (coning the saltwater into the wellbore with sand).

4. High-producing wells not only result in high water production but produce a considerable volume of sand, silt, clay, mud, suspend formation scales, and crushed shale. (Thompsons, Texas; Goose Creek fields near Baytown, Texas; the Gulf Coast wells; and shallow wells of west Texas.)

5. The following fluid lifting methods may produce a high volume of fluid with formation solids:

	Fluid Lift Range (bbl./day)	
gas lift methods	100–8,000	barrels/day
electric submersible pumps	100–9,000	barrels/day
hydraulic lift pumps	100–6,000	barrels/day
hydraulic jet pumps	100–5,000	barrels/day
PCP pumps (rotor/stator)	400–4,000	barrels/day
beam tubing pumps	5–3,000	barrels/day

The fluid lifting capacity of the above-listed methods is measured based on various reservoirs with unconsolidated formation sand and solids. The tight formation may yield fluid differently.

None of the above artificial lifting methods can function properly if the wells are producing sand, mud, shale, and scales.

A high volume of sand abrasion and fluid production will reduce artificial lift performance and the useful life of downhole equipment.

All the natural-flow wells and the artificial-lifting wells will practically produce certain amounts of sand and solids regardless of preventive measures.

Higher oil-cut reservoirs produce minor volume of formation sand (oil-wet reservoirs).

As the water production increases, the sand, solids, and scale will increase. High-yield water production wells will produce more sand.

Washing and cleaning flowing sand out of wellbores and the production equipment repair are costly facts to oil-producing companies.

Produced sand will build up in the tubing string, the surface flow lines, the production equipment, and the bottom of stock tanks and will cause equipment failure, bad oil, and well equipment repairs.

High-rate water-drive reservoirs may yield high sand production.

Reservoir structural damage and economical impact associated with sand production in the oil-and-gas wells cannot be ignored.

Higher oil-and-gas production rates out of horizontal wells in the beginning seem to be exciting; however, after several months, major remedial workovers and high cleaning costs will hit the books!

Mass production of unconsolidated formation sand of Gulf Coast wells offshore and onshore is an alarming concern. Producing two hundred thousand barrels of saltwater per twenty-four hours in one single field containing formation sand is an alarming problem (environmental and operation cost concerns).

High production without gravel packing may cause numerous damages:

- reservoir permeability damage because of rearranged formation-sand grains
- sanding out wellbore and flow lines
- create erosion and sanding out rods and pumps
- sanding out production tubing and packers
- sanding out surface and subsurface production equipment
- bad oil and tank battery because of migration of sand and sludge
- higher cost of sand transportation and disposal of salty sand and sludge
- casing collapse and doglegs because of large cavities behind casing
- collapse of reservoir formation behind gun-perforated casing

- subsidence, sinking of formation, and collapsed casings
- wellbore subsidence, fishing, and casing corrections
- saltwater disposal facility cleaning

The breakdown of calcareous formation bonding in some semi-consolidated reservoirs formation may be some of the causes of abnormal sand production.

The key to successful wellbore sand control is an early detection and quick reaction to stop reservoir permeability and porosity damages.

Unconsolidated sand plays an important rule of supporting shale and overburden matrix from mechanical breakdown. (Do not disturb reservoir formation when gravel packing!)

Rearranging the reservoir formation sand may reduce porosity and permeability.

Further delay in sand control will cause the formation wall to cave into the wellbore, causing skin damage and costly fishing, and make sand control less successful.

Control the Unconsolidated Formation Sand by Gravel Packing

The author's gravel-packing publishing material and comments is based on forty-one years of practical remedial and workover experience, reservoir sampling, gravel packing, reservoir fracturing onshore and offshore of south Louisiana, Mississippi, Texas gulf coast, and New Mexico, USA, while working for Gulf Oil, Chevron, Pennzoil, Devon Energy, and AmeriCo Energy Resources.

The author wishes to present the materials based on the simple practical facts and field experience of the past and present without boring formulas, charts, and graphics.

There are two major methods of gravel-packing sand controls.

I. Standard gravel-packing method

    a. Run a selected sized screen and liner into a well and pump gravel-packing sand around the screen and liner.

    b. Prepack sand-gravel into the formation and perforation tunnels, then run and set the screen and liner in place across the open perforation and pump sand-gravel around the screen and liner as much as practically possible.

       The prepacking is the best technical approach to obtain tight gravel packing through the perforated tunnels and formation.

II. Conventional well-completion sand-control method

       Run and set a selected sized screen and liner and complete the well without gravel-packing sand. (This method is done based on the reservoirs naturally produced coarse sand grains from the well to pack and cover around the screen without additional gravel-packing sand.)

       Gravel packing is an art of controlling formation sand before getting into the wellbore.

The Standard Gravel-Packing Sand Control Techniques in the Oil Industry

There are basically two mechanical/hydraulic approach of carrying the gravel-packing sand into a well:

- ► By using nonviscous fluid application (water-pack method)

- ► By using high-viscosity fluid application (high-viscosity slurry-packing)

Nonviscous Fluid Application

The mechanical gravel pack using plain, clear filtered water is known as water-pack method. Clear filtered water will only be used in gravel-packing process as carrying fluid.

There are several water-based gravel-packing techniques:

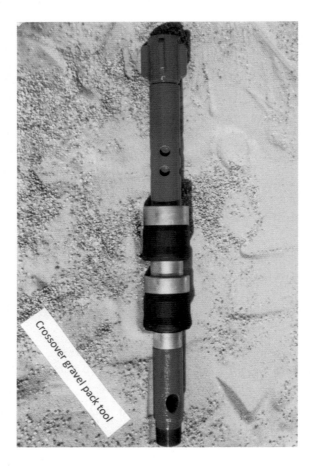

Crossover gravel pack tool

    A. Poor-boy gravel packing (reverse circulating method)
    B. Cased-hole prepacking sand control using crossover tools
    C. Cased-hole prepacking sand control using wash-down method
    D. Through-tubing gravel-packing methods
    E. Open-hole gravel-packing methods in oil-and-gas wells

High-Viscosity Fluid Application

Mechanical gravel-packing method using crossed-linked or viscous polymer-gelled fluid is referred as high-viscosity slurry-packing. Gel-based fluid will be the primary fluid to carry the sand slurry during the viscous slurry gravel packing. Slurry gravel-packing may be conducted similar to fracture-packing with high sand concentration and higher hydraulic pump pressure application.

We will present each of the above gravel-packing techniques in the following pages.

Liquid resin and resin-coated sand are often used to fracture and/or gravel pack a well.
The liquid resin sand-control method is tested to be the least effective gravel-packing method.
The resin-coated sand should be used under ideal wellbore conditions.

I applied liquid resin and the resin-coated sand in four different wells. The liquid resin and resin-coated sand chocked off the reservoir production (results were not too impressive).

See pictures of liquid resin and the resin-coated gravel-packing sand.

Caution to the Well Operators

Some wellbores may be sensitive to the following:

- water-based gravel packing
- gel-slurry gravel packing
- liquid resin or resin-coated sand-control packing

In order to stimulate or gravel pack sensitive reservoirs, you must use non-asphaltic hydrocarbon and/or a similar compatible light oil product or non-explosive chemical of high-quality foaming agents in the well treatments.

Never use dry air or oxygen in air-foam units in the oil-and-gas wellbore cleaning without adequate water mixture. This may cause explosion, based on two major incidents.

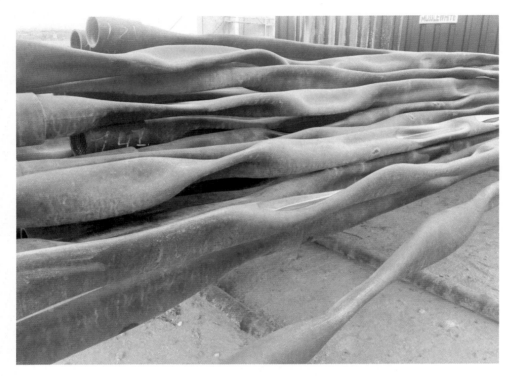

****

The Water-Pack Sand-Control Techniques

In the water-pack sand-control procedure, the carrier fluid is only clean and clear, filtered produced saltwater. The proponents of water-pack gravel packing believes that there is nothing better, cheaper, safer, more reliable, and readily available gravel-packing method than clean, thin, filtered water-pack sand-control method—a worry-free sand-control technique in the oil patch.

The following are advantages of plain filtered water-pack gravel packing:

- Produced water is compatible fluid to most reservoir formations.
- It is cost-effective—lower cost than slurry packing.
- There are readily available low-cost material and carrying fluid.
- It is safer to work with, without toxic chemical additives, and is nonhazardous.
- It is a tight gravel-packing process.
- It is free from cross-linked gel residue and honeycomb solids.
- There are no sand-bridging problems during gravel packing.
- There is no sand crushing during gravel packing.
- The work procedure techniques are dependable and simple to follow.
- It produces dependable gravel-packing results (if it is done correctly).
- It has longer and useful life expectancy than the gelled slurry packing.
- It has no formation disturbance (no formation damage).
- It is successful in nine hundred to ten thousand wellbore operating range.

The following are disadvantages of water-pack gravel packing:

- It has lower sand concentration compared to slurry packing.
- It uses more water compared to slurry packing.

The low-fluid pack velocity and controllable low sand concentration with basically no residue will make the water-based gravel packing very unique and favorable among the supporters of water sand packing in the oil field, especially in the low bottom-hole pressure, low-production wellbores in the Gulf Coast area.

Successful gravel packing of one thousand to eight thousand pounds of sand can be achieved in some reservoir formation using low-density resieved gravel and plain, thin, filtered produced water (Thompson's field, Goosecreek field of Baytown, TX. and South Louisiana–Quarantine Bay fields).

Compatible thin fluid and low gravel concentration in the water pack is used as natural fluid loss or bridging material with tight gravel compaction across the perforation and behind the casing. The artificial chemical and loss circulation bridging material will not be used in the water-pack sand-control techniques.

Sand pack do not frac-pack any unconsolidated reservoir formation.

The thin freshwater sand-packing technique is not a high-sand concentration carrier because of the low viscosity of freshwater. If reservoir formation takes water, it will take sand and gravel as well.

The water-based gravel packing method is based on low pressure and low-sand concentration sand packing in order to Braden-head the maximum volume of gravel through the gun-perforated holes and into the formation without fracturing or crushing the bonding structure of reservoir formation.

In nonviscous fluid (water pack), one half to three pounds of gravel can be carried out with a pumping rate of two to three barrels per minutes. Tight gravel-packing sand can be achieved through the open perforation and onto the formation because of the low-viscosity fluid.

Gravel-Packing Sand Placement Techniques

In a water-pack gravel-packing method, a simple, special gravel-packing unit will be used to mix the sand with water and pump the mixture into the well. A small and simple-designed sand hopper drum is used to mix, pump, and displace the gravel-packing sand without high-speed mixing blades to avoid crushing, grinding, or polishing the sand grains. The unit is capable of delivering three hundred or four hundred pounds of sand at a time on each stage without bridging problems (a reliable and efficient technique to mix and deliver the sand).

Each stage of sand packing is controlled by mixing and displacing three or four pounds of gravel per gallon of sand slurry in plain, thin, filtered water/sand slurry without bridging.

One thousand pounds to eight thousand pounds of resieved sand gravel can be pumped and packed into a formation and around a screen and liner with great success based on the formation characteristics.

Some formation may take one thousand five hundred pounds of prepacking sand, and some reservoir may take as much as eight thousand pounds or more prepacked sand behind the casing without any slick, gelled polymer carrier fluid.

In a tight shale-and-clay content reservoir formation, the chances of placing large volumes of sand into and across a formation is much less. High-pressure gravel packing will force the sand into the formation containing the clay bed and may block off the permeable zone (the high pressure will pack off the clay or shale strata tighter).

The high porosity and permeable reservoir formation will take large quantities of gravel-packing sand without higher hydraulic horsepower injection pressure. If a reservoir is taking water, it will take sand-gravel without high injection pressure. At low sand concentration, one can pump or Braden-head-pack a large volume of high-quality resieved gravel behind the perforations (tight gravel-packing concept).

Water-based gravel packing is a simple method with full controllable packing method, and it is less expensive than high-cost slurry packing.

****

Success or failure of a sand-control completion depends on the following factors:

1. Quality and quantity of gravel-packing sand

   Gravel-packing sand is derived from crushed, reprocessed manmade silica gravel. The quality of gravel-packing sand is based on the roundness consistency, crush resistance, and/or acid chemical solubility (impure gravel will break down with acid).

2. Method of gravel-packing placement to obtain tight sand pack

3. The reservoir's rock characteristics (porosity, permeability, shale, and clay content)

4. Wellbore preparation—casing, tubing, and borehole cleaning is of utmost importance to avoid contaminated mud, sand, and paraffin solids into the formation.

5. Cement bonding behind the production string

   Cement is one of the important and tangible materials in any oil-and-gas well. Cement may be the cheapest in oil-and-gas operations; however, the importance of good cement bonding around the casings cannot been emphasized or expressed enough.

   Lack of cement bonding will expose the oil-and-gas reservoir production to all types of thief zones in the wellbore and will leave the casing string without protection, causing casing failure, unwanted high-yield water production, and caving in of loose formation sand and solids into the wellbore.

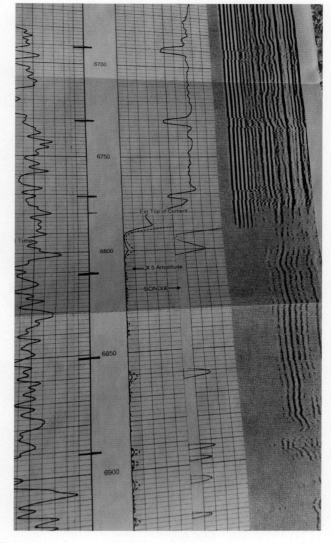

6. The production casing's mechanical integrity (MIT)

   One must test and repair the casing string before gravel packing a well. Squeeze off and test the casing for leaks before doing any gravel-packing work to avoid gravel-packing failures. A clean casing without leaks, pipe scale, and rust is very important in the outcome of any good-quality gravel packing.

7. Perforating holes during completion

   a. Underbalanced perforating is preferred. The advantage of underbalanced perforating is to remove sticky mud filters and viscous gel from the face of formation. Tiny formation pores must be kept clean and free from gel pills and drilling mud material.

   b. Perforating holes of six shots to eight holes per foot with entry holes of 0.3″ to 0.50″ is preferred before gravel packing. (This will reduce differential pressure.)

   c. If a well is perforated through tubing, one may consider to reperforate with a casing gun or conveyed guns (TCP).

   d. High-density perforation may reduce differential pressure across the open perforations and avoid fluid blasting.

e. Exercise caution when perforating through old casing strings to avoid large splits or parting the casing string because of the high power of explosive blasting.

f. Use hollow-carrier casing guns to avoid gun junks into the wellbore and beyond the casing (do not use slick guns).

g. Avoid using zero-phasing perforating.

In the zero-phasing, gun-perforating practice and procedure, the perforating gun will be loaded to shoot in one direction only as guided by applied charged magnet.

Zero phasing means "all the shots will be fired on one side of the casing only."

Through-tubing zero phasing is normally conducted in large casings with a smaller tubing string (zero phasing makes the casing string weaker at the gun-perforated holes and may cause shifting with time).

90 degrees or 120 degree perforating phasing is preferred in gravel packing.

h. Collapsed casing, parted casing, doglegs, tight spots, and horizontal wellbores. (This will be discussed in the next chapter.)

The gravel-packing companies should not be responsible for a gravel-packing failure as the result of unknown casing mechanical problems!

The casing string must be tested and drifted to make sure the screen and liner can successfully be run and/or pulled from a wellbore to prevent costly fishing work and to avoid damage to the screen and liner.

An unwrapped stainless screen will cause the gravel packing to fail and will be difficult to fish, and it may loosen the reservoir interval.

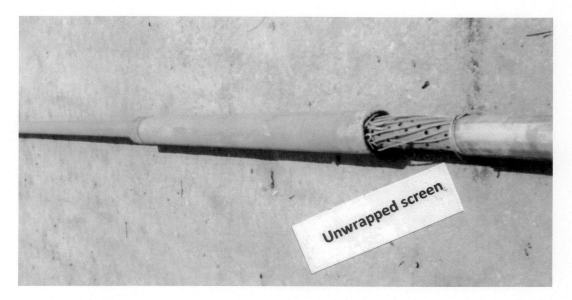

Never force the screen and liner through casing tight spots. (Jarring, turning, and twisting a new screen to get through tight spots in a casing is a bad idea, and it is an unprofessional practice.)

You may force the screen through, but you do not know what you actually damaged!

The casing must fully be tested and drifted before building the screen and liner assembly.

(If you build it, you bought it, and you may not be able to run it!)

Special screen and liners can be designed to run through tight spots and deviated boreholes with great success if necessary. (Do your homework on the screen selection in advance.)

8. Holes in the casing string

Holes in the casing string are one of the major causes of gravel-packing failure. Casing holes must be squeezed off and tested before gravel packing. Do not gravel pack any well without testing and inspecting the casing string.

Never conduct gravel packing in a wellbore with shallow casing holes or leaks.

Shallow casing holes above a screen and liner will cause the gravel pack to fail by dumping sand and solids into the new screen and liner as soon as you put the well in production.

(Your gravel pack may fail the next day with a big surprise!)

Remedial workover in sanded-up gravel-packing screen and liner will be a costly project. It will cost several times more than the initial gravel packing.

Failed gravel packing in the offshore operation is a major cost issue. (It is a disaster that you do not know how to explain—your gravel-packing failure—to your boss!)

Casing Leaks and Splits

- A casing leak will cause the casing to become parted with time. (You may not be able the pull the equipment below a parted casing!)

- A casing leak across a wet reservoir sand may cause unwanted high-water production and will cause oil-and-gas reduction.

- Casing splits can successfully be repaired using casing patch and cement.

- Casing leaks will dump sand and iron sulfide bacteria on the production equipment, and they will plug off the screen and liner.

- It is an expensive scenario to operate a well with casing holes.

- A casing hole near or just above an open perforation can be repaired by using

    a. cement squeezing or

    b. running and placing the screen across the hole and gravel packing it in place.

    Never gravel pack a well with casing holes above the screen and liner.
    (Don't say you did not know!)

****

Wellbore Preparation before Gravel Packing

Never take shortcuts on the wellbore preparation to save cost! Cost cutting is an obstacle to prevent you from doing successful work. (It might cost you several folds later!)

The purpose of the bridge plug (footing) in a cased hole well is to support the screen and liner. A cast-iron bridge plug may be run and set below the gun-perforated holes to set the screen and liner on a hard spot in the casing.

A bridge plug is also used to isolate the open perforations from the wellbore below and also to drift the casing string before running a screen/liner.

A cast-iron bridge plug or a solid-cement bottom is necessary to isolate and to support the screen and liner assembly along with suspended gravel-packing sand in the annulus around the screen and liner assembly (keeping the sand from falling into the rathole).

A production isolation packer with a stinger assembly may be used for the same purpose.

The advantage of a stinger-type packer below a gravel-packing screen and liner is to allow the mud, silt, and fine-grained sands to fall into the rathole (using rathole as a mud anchor).

It is an option to link or connect into different gravel-packed zones below.

Below the open perforations must properly be isolated to support the screen and liner with gravel sand and also to avoid leaks and thief zones from the bottom (avoid crossflow and channeling).

The bridge plug or foot base should be set in the casing string not more than twelve feet below the bottom set of the open perforations. (Do not set the plug too close to perforations!)

If the bridge plug (footing) is set too deep, the following corrections may be needed:

1. You may drill out the existing bridge plug and set another bridge plug higher.
2. You may set another bridge plug above the existing plug higher.
3. Extend a blank liner below the screen section to reach the plug (for only a short distance not deeper than fifteen feet).
4. Run and set an isolation packer below the screen section, if necessary.
5. Fill up the short void space with sand, and cap the sand laden with cement to obtain a correct distance with the hard base as required.

I do not recommend to set any screen and liner on top of a long interval of sand bridge because of possible voids or "holiday" in the sand bridging. (It might disappoint you a great deal later.)

If you have a formation-sand bridge that you can use, you may dump bail fifteen feet of cement on top of the sand before using it as a dependable plug.

Several gravel-packing jobs have failed as the result of bad suggestions on setting the screen/liner on top of a long interval of sand bridge to save cost! (Do not try it!)

The footing under the screen and liner must be hard, stable, and sealed off from the bottom to avoid the screen and liner from sinking down the hole and prevent the gravel-packing sand from falling through the rathole below the open perforations.

A misplaced or sinking screen and liner will block off the open-production perforations and will be costly to correct or fish out. (Always tag the screen and liner after gravel packing to make sure the job is done correctly.) Sinking gravel-packing screen and liner will block off the oil production, and it is costly to correct.

When the screen and liner sinks, the blank-liner will block off the open perforation. The entire gravel-packing assembly and the sand will be misplaced and out of order. The well will not produce the expected production.

****

Oil Well Screen/Liner and Related Setting Tools

The art of gravel packing and application of sand screens in the oil wells!

All the components of the screen and liner assembly must be designed using a good and dependable quality material to extend the useful life of a gravel-packing work.

The components of the screen and liner may be subject to change, depending on the gravel-packing method, wellbore condition, and the application of the setting tools.

The following parts may be used with the application of a screen/liner in the gravel packing:

• **Bull plug:** A blind-end plug with external or internal threads is used at the bottom of the screen/liner section to land on the base plug. A bull plug is used to avoid sand or solids entering the screen from the bottom. Blind bull plugs are used when there is no need to wash the screen and liner in place!

(There are several types of bull plugs to select from.)

Bull plugs with welded blade may be used at the bottom of the screen section as a friction-drag tool to prevent the screen and liner from turning when backing off from the liner assembly. Do not use the bull plug to cut or drill solids deeper!

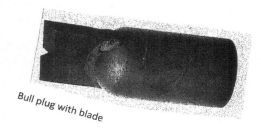

Bull plug with blade

Round-nosed/or flat-bottom bull plug with welded flat-bottom steel plate is used to create a wider support and prevent the screen from sinking into the sand bridge below the screen.

Plugs may be made with brass, bronze, or simple carbon steel material.

Brass or bronze plugs will last longer and resist corrosion longer than carbon steel!

• **Set shoe:** The wash-down shoe has an application similar to the bull plug, with exceptions. A wash-down set shoe is screwed at the bottom of the screen section to wash down the screen and liner in place going through sand gravel.

A set shoe is spring-loaded with a ball-and-seat-type check valves at a closed position and will open with fluid pressure, allowing clear fluid to pass through while washing the tool through sand gravel.

When the washing-down operation is completed, the shoe will automatically close shut with spring force to prevent solids from backflowing into the screen from the bottom.

There are several types of wash-down shoes available. Shoes are used based on practical applications.

The turbo-type jet is also available for the same application purpose. A set shoe with turbo blades are often used in washing the screen and liner in place. A jet-type set shoe will accelerate washing, and it reduces critical washing time (turbo blades).

• **Perforated inner jacket of the screen section:** The inner jacket of a screen may be referred as base-pipe (an important part of the screen that supports the screen and allows the production fluid to pass through). The base-pipe is made from plain, slick carbon steel liner tubing of H-40, J-55, or L-80 pipe, or other choices with precision machine–drilled circular holes.

- **Inner jacket circular perforated holes:** The perforated holes on the screen's base-jacket is necessary to allow the production fluid to flow through with ease.

  Several circular perforated holes per foot are drilled through the pipe with accurate precision control and deburred internally as well as externally to remove sharp slags. Perforated holes are drilled accurately to furnish the required inlet area at a unique pattern. The number of perforated holes per foot on the base-pipe depends on the pipe size or the customer's request (holes may become enlarged and washed out with time).

  Heavy wall, full-opening blast joints should be used as the base-pipe to support and protect the perforated holes from corrosion and fluid blasting.

  The liner jacket may be internally and externally coated before screen wraps as per customer request to slow down corrosion rate and to extend the useful life of screen and liner.

  Corrosive environment and high temperature must be considered when designing the inner jacket of the screen.

  Stainless base-pipe or chrome tubulars maybe used in the manufacturing process because of corrosive H_2S or CO_2 environment.

  Carbon steel base jackets will corrode with time in the corrosive environment and may cause stainless screen wraps without sufficient support.

- I recommend using full-drift blast joint tubulars for screen jackets to slow down the rate of corrosion and tubing collapse.

The Vertical Pipe Slots on the Base-Pipes (the Inner Jacket)

Slotted liners are often called bare-sand screen (no screen). Bare-slotted liners were actually the first devices used in gravel packing to restrict and support the formation sand in the open-hole sand control.

Slotted liners are widely used because they are the most economical tools, especially in long-formation intervals. Slotted liners may be economically justified to some people, but they are not the most effective alternative source in modern gravel-packing methods.

Square vertical slots are often used on the base-pipe for water-well screens and open-hole gravel-packing practices. Slots are used in wire-wrapped screen in oil-well gravel packing as well.

Pipe slots can be made in several form and patterns:

- staggered rows
- horizontal slotted shape
- nonstaggered rows
- gang-slotting staggered rows

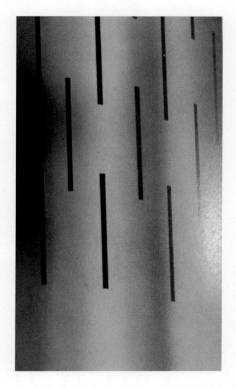

Pipe slots may be machine cut either straight-sided or key-shaped.

Vertical slots are precision-machine cut in staggered rows and deburred similar to circular perforated holes to allow large fluid flow.

Vertical slot screen is suitable for water wells and open-hole oil wells with coarse formation-sand gravel. Slotted pipe can be used with or without wrapped screen.

Slot opening may range from 0.02″ (1/50″) to 0.25″ (1/4″) on various pipe sizes. The number of cut slots on a base-pipe may depend on the size of the base-pipe (ranging from one-inch pipe to sixteen inches outside diameter pipes).

Never cut slots with a cutting torch; they seldom cut accurately!

Slotted pipes have less resistance against corrosion and fluid erosion.

Slots will become larger with time because of fluid and solid particles blasting through the circular holes.

The design criteria used for bare-slotted pipes are as follows:

> For non-uniform formation that is poorly sorted sand
> $W = 2D10$ (inches) W = the width of the slot
> $D10$ = the percentile point
> For uniform formation-sand sample that is well sorted
> $W = D10$ (inches)

****

All-Welded and Ribbed-Welded Wire Screen Selections

- The building and manufacturing of all-welded screens are unique and require precision engineering work. All-welded screens are stronger and built more accurately.

- All-welded screens are built tougher and stronger than the standard screen. All-welded screens are made to resist deformation during the running, washing over, and fishing operation without unwrapping problems.

- The V-shaped stainless-steel wires are precisely wrapped and firmly welded to wire-ribbed rods at each contact point to protect wire gaps and to prevent twisting or unwrapping.
No external welding beads and/or solder welds are used on the all-welded screens.
(All-welded screens are the best-manufactured screen available in the market.)

• The stainless screen is built from V-shaped wire wraps on special stainless wire ribs at a constant tension, controlled speed, and precisely measured, spaced slot gaps.

— Gaps are gauged based on selected gravel-packing sand.
— Gravel-packing sand is selected based on the formation-sand analysis.

• Depending on the size of the screen, several wire ribs are used to support the V-screen wraps in place (twenty steel ribs are spaced uniquely to build 2-⅜″ screen and twenty-six wire ribs are used to make 2-⅞″ screen).

• Stainless steel screens are made of tough materials and will resist against corrosion, fluid erosion, and deformation. It is difficult to mill over or fish unwrapped stainless wire screens.

• Stainless screen wires are specially made of 304 or 316 stainless steel or Hastelloy steel wire material.

• Stainless wires are spirally wrapped over and around vertical steel rib rods and welded at each contact point with precise control (the tip of the V on the rib).

Stainless wires are extremely tough to mill and can resist against high corrosion, fluid erosion, and deformation at high temperature at twelve-thousand-feet deep holes.

- The vertical wire ribbed-type screens are built and cut to precise length. The screen will be checked for quality control, slots are gauged, and the screen will be slid over the perforated base jacket like a sleeve, pressed, and welded on both ends of the base-pipe without blocking the perforated holes.

- Stainless rod ribs (base rods) are made from the same material as wire-wrapped screens. Ribs are spaced with accurate precision design before welding. The purpose of vertical rod ribs in the screen is for the space and strength support.

- Pipe-base ends are used on some screens, threaded, and coupled on various sizes.

- Wire ribs are precisely spaced so that the ribs will not block off or cause restriction at the face of inlet/ outlet perforating holes on the base-pipe.

- The accurate design and selection of wire-wrapped slot gauge are extremely important to the success or failure of sand control.

- Gravel-packing sand selection is based on the formation gravel. The wire-wrapped slot gauge is based on the gravel-packing sand selection.

- Gravel-packing customers should have a basic knowledge about screen-and-liner manufacturing process to appreciate the art of screen making and to make intelligent decisions on selecting correct-sized screen and liners.

- Wire-wrapped gap and slot gauge are of utmost importance in screen design. The slots must be designed, spaced, and gauged based on the true, analyzed formation-sand grain representative.

- Reservoir sand-sample analysis is vital before determining the screen gauge. Screen-gauge slots should be built based on true, matching formation sand with gravel-packing sand size. (Example: 20/40 gravel-sand versus the formation sand and 12 Ga = 0.012 gauge screen)

- Dry formation-sand samples must be analyzed to determine the correct size of gravel-packing sand screen for a particular well before making up the screen and liner.
  Large screen gaps will allow the fine formation sand to pass through, and small screen space- wrapped gaps will choke the well productivity and may plug off the screen.
  (Do not build the screen until you obtain sound sand representative samples!)

  ****

- Reused screens and liners are not recommended for any wellbore gravel packing!

- Reused screens are difficult to pressure wash and may not gauge properly because of fishing operation or mishandling (twisting, stretching, and jarring).

- Do not build any screen and liner before drifting the casing string and the open perforations! (Do not jump ahead. Once you build it, you will own it.)

- Stainless screen slots will be spaced based on gravel-pack sand-grain size to prevent the smallest formation-sand grains from passing through the sand pack.

Selection of Gravel-Packing Sand

- The selected resieved gravel-packing sand may be six to eight times larger than the smallest resieved grain of the formation-sand samples taken from a well's reservoir formation (ten percentile points may be noted on some cases).

- Local practical experience may be applied in some gravel-packing cases. Some formation-grain sizes are very small and look like fine baby powder.

- The screen and liner's outside diameter must be small enough in order to place a larger volume of gravel-packing sand between the casing and the screen and liner.

- The larger the screen and liner size are, the lesser is the annulus space for gravel-packing sand.

    The screen/liner must be designed based on the casing's inside diameter.
    Larger casing string will have better options of using larger-sized screen and liner.

    Do not force a large-sized screen/liner into a small-sized casing, as there will not be enough void space for the gravel pack to hold the formation-sand pack.

- Gravel-packing sand is subject to disappear or blast away with time because of the fluid's turbulent movements (gravel-packing height and space are important).

- When there is not enough space in the annulus, it is difficult to wash over or fish the screen and liner out without unwrapping the steel wires.

****

Selection of Screen/Liners Based on a Wellbore's Condition

The screen and liner may be designed and constructed based on the wellbore's physical condition.
The older casing wellbores may have various physical obstructions that must be dealt with before building a screen and liner.

Casing conditions such as tight spot, no cement bonding, leaks, holes, scales, or shifted pipe at perforations must be evaluated before building the screens.

Casing inspection, formation-sand sampling, and the wellbore's integrity are the duties of a well owner before ordering a sand screen.

****

Types of Manufactured Gravel-Packing Screen and Liners

Screens and liners may be built or reshaped based on downhole parameters.

- Standard screen and liner: This consists of inner perforated tube jacket and wrapped stainless screen (all-welded screen or rod-type screen). The standard screen is applicable for all the normal wellbore conditions.

- Prepacked sand screen: This consists of an outer screen section, resieved-sized gravel, and the inner screen section with a base-tube jacket (gravel sand is packed between the two screen sleeves and welded from top and bottom of the base-pipe).

  Prepacked screens may be built and used in all oil, gas, or saltwater application without further formation prepacking (prepacked screens have larger outside diameter than standard screens).

  Prepacked screens may be built in various designed configurations. I consider a prepacked screen as a band-aid idea between a bare-slotted liner and wire-wrapped screen without gravel packing.

  In my evaluation, the life span of a prepacked screen in oil-and-gas wells is shortcoming. The formation fluid and sand tend to blast away the prepacked sand content and/or will form a bridge by fine reservoir unconsolidated sand.

  I do not recommend the use of a prepacked screen in any oil or gas wells as a sand-control mechanism. (They will plug off with formation sand.)

  The application of prepacked screens with additional formation prepacking may reduce the injection rate in some saltwater disposal wells and require washing with low concentration of acid treatment and/or remedial work.

- Inner screen with slick outer perforated pipe: Consists of wire-wrapped screens that are especially built inside of a perforated tubing jacket. This type of screen and liner is designed to be used in some wellbores with mechanical casing problems and/or tight restriction through the perforations.

- The slick pipe jacket may be designed with vertical slots or round perforated holes. This type of screen and liner can be run in and out of tight spots without damaging or unwrapping the internal screen section. (It is not a standard-made screen.) The purpose of this type of screen and liner is to avoid unwrapping or damaging the screen by forcing the screen/liner through a tight spot or shifted gun-perforated casing area.
  Particular attention must be directed to perforated slots designed to prevent slowing down the rate of production through the screen.

- Blank pipes (blank liners): Blank liners are made of a plain low carbon steel tubing of J-55 or H-40 and extend from the top of the screen section up the hole as long as needed.

  The application of N80 or L80 tubing is at the customer's discretion. Upset or non-upset plain pipe can be used in building the blank liners.

  H-40 and/or J-55 are made of low carbon steel with lower tensile strength and fair corrosion resistance.

  The L80 and N80 and P110 pipes are made with higher tensile strength and less corrosion resistance because of higher carbon content. The relationship of carbon and steel is just like freshwater and salt!

  Blank liners are plain pipes (no holes) designed to hold and support the screen section and keep the gravel-packed sand around and above the screen and liner in the annulus above the perforations (the longer liners will hold large volumes of sand in the annulus above the screen).

- In 1993, I ran a 450-feet screen and liner across four zones and prepacked in place with success (Goose Creek Bay).

- The blank liners are cut and machined from thirty- to thirty-two feet long per joint of pipe and threaded on both ends. The blank liners need to be internally and externally coated to resist corrosion and erosion and to last longer. (Liners may be threaded with 10v, 8rd, and/or square buttress-type threads.) Buttress threads and eight-round threads are stronger than ten-round, 10v, or integral thread connections. (Correct makeup torque must be applied properly.)

- The centralizers: The blank liners and the screen section must be centralized going through a casing string and/or open-hole completion. Guide centralizers are normally spaced and welded on the screen and liners. The bow-spring centralizers are normally used in the open-hole gravel packing. (This has the advantage of expanding further.)

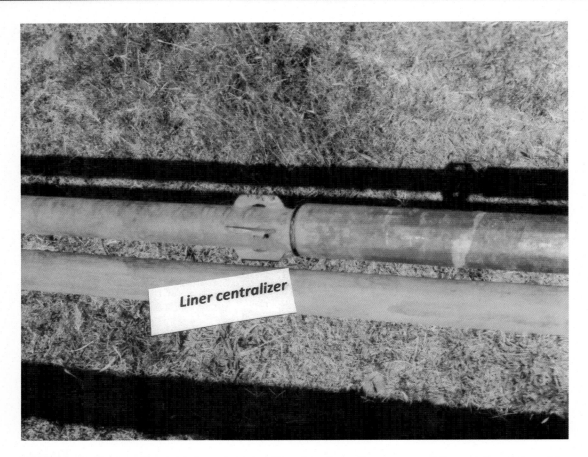

Liner centralizer

- The purpose of centralizers on the screen and liner is to keep the screen and liner at the center of the wellbore and to allow a uniform sand-slurry distribution around the screen and liner assembly.

- Centralizers will prevent bridging on the high side and/or the low side of a borehole in the deviated holes, doglegs, and tight spots.

  There is no such a thing as straight-hole drilling in the oil-and-gas wells, only controlled drilling. (Some holes are drilled like a moving snake!)

  After a hole is drilled, you may be faced with crooked and deviated holes of many kinds.

  The reason for having so many crooked holes in the oil field is the time versus cost. Crooked holes cost oil companies a great deal of money and equipment failures. (Holes in tubing strings and rod wears are some of the results of fast-drilling crooked holes.)

  For your information, a three-degree hole from vertical is considered as straight of a hole in many oil or gas wells of today.

  The centralizers on the blank pipes must be built to last longer and resist corrosion and break apart with ease during washing over and/or fishing.

  The bow-spring centralizers, bolted aluminum, and plastic centralizer are not stable. When washing over the fish. The screws will break apart because of corrosion and cause drag and torque during fishing and retrieving the screen and liner (floating fish). Steel bow-spring will not last in corrosive wellbores.

- The recommended space between the centralizers is about twelve to fifteen feet apart.
  Do not carry out with overloading the screen and liner with too many centralizers.

- Centralizers are the largest diameter on the screen and liner assembly and closer to the casing wall. (Often you may have to grind down the tip of some centralizers going through the tight casing spots or gun-perforated holes.)

- Running a gauging ring through the casing perforations to the top of bridge plug may save you time from going in and pulling back out with the screen.

- When you design a screen and liner to be run in a well, always ask yourself, How can I run and pull it back out of the well?

  The casing wall across the gun-perforation interval in the old wellbore often appears to be washed out or shifted because of corrosion/erosion and often forced inward (deformed perforations).

- Avoid welding the centralizers at the face of the screen section and directly across from the open perforations. This will cause a hole and create a potential screen failure.

- Any exposed blank-pipe section across open perforations will fail with time.

- Centralizers are either made of aluminum or thin-walled piece of carbon steel metal sheet, cut and welded to the screen and liner joints in semicircular shape (cut washer).

- The aluminum centralizers are easy to drill; however, the screws will break and cause floating fish.

- Bow-spring centralizers are used for cementing and/or wireline work or open-hole gravel packing only (space at forty feet apart or as needed).

- Spring centralizers will corrode away with time and become parted during the washing over and will create floating junk in the annulus and over the screen.

- The carbon steel centralizers are welded. Welding on the blank liner and the screen jacket may degrade the pipe's structural integrity. (Do not hot weld!)
  Do not cool off the welded pass area with water. Always wait until the affected area cools off with the air. The application of heavy epoxy coating is recommended on all the welding or scratched spots on the screen and liner.

- Centralizers are cut and welded to screen and liners in semicircular and/or a round shape with minimum contact point in order for the sand to pass and/or to wash over the screen and liner easily using wash pipes and the wash-over shoes.

- I prefer semicircular washer-type centralizer with less area of contact to casing. They fold and cut easy during washing over.

- The centralizer must be ¼″ smaller than the measured casing drift, passing through the perforations to avoid complication.

****

The Formation-Sand Sieve Analysis

The importance of sand samples from a well must not be taken lightly. The screen and liner must be built based on the well's sand-sample analysis. Without sand-sample analysis, your gravel packing could fail (sand samples could be different from one well to another well in the same field).

Sand samples will be taken from a wellbore by the following:

► A full-sized core during drilling operations—a full-sized core sample is the best choice and can be used for gravel-packing formations.

► Side-wall cores—several side-wall cores are necessary to obtain true formation sample.

► Wireline sand bailers—these samples are taken from the rathole at the bottom of perforations with large sand grains. Wireline sand bailer sampling is questionable and seldom representative.

It is difficult to obtain representative sand samples from perforated cased-hole wells.

► Circulated samples during a workover operation must be taken by using a pair of socks and not a bottle.

► A pair of socks or a five-gallon bucket may be used to catch fine and coarse sand grain samples while circulating the solids at low pumping rate to the surface.

► Sample taken from the bottom of a mud anchor joint from a pumping well is a more reliable sample. (Do not back off the bull plug until the water is drained. Avoid splashing the sample.)

► Sand samples may also be taken from a wellbore by using a sand bailer and swab-line (sand line).

> ► Samples taken from the production vessels, such as separators and heater treaters, are often used in sand analysis. These samples may come from several wells and are seldom the correct samples.

The open-hole wire-line plug sand sample or full-core sample can be used if available. The reservoir formation may not be consistent in regard to the shape and size of sand grains throughout the perforating interval.

Shale, sand, and clay strikes may be spread across the interest zone, and it may be difficult to catch representative formation samples.

Coring every five feet of reservoir depth may be considered as a matching sample.

Successful local sand representative sampling may be used in some cases before building a screen and liner.

The sand sample can be different from one zone to another zone within the same wellbore

****

Selection and Evaluation of Gravel-Packing Sand

After the selection of an acceptable formation-sand sample, the screen and liner and gravel-packing sand will be selected.

The success and productivity of an oil well after gravel packing is based on the accurate screen analysis. The quality and quantity of gravel-packing sand grains and method of placement of sand-gravel through the open perforation is very important.

Sand-gravel is the first filtering block that holds the formation sand in place. The steel screen is the second filtering block to protect the sand-gravel and permit the well fluid to pass through.

Gravel-packing sand must be thoroughly checked for grain size and shape. The gravel-packing sand and the fracturing sand is the reprocessed manmade silica sand-gravel.

The gravel-packing sand is washed, dried, and screened to remove impurities and particles before shipment. Gravel-packing sand is selected based on the upper limit of sand-sized distribution.

One should not use fracturing-gravel sand in the gravel-packing process. (It contains too much crushed sand and particles.)

Roundness consistency, crush resistance, and chemical acid solubility are the major requirements for gravel-packing sand.

High-concentration acids such as HF (hydrofluoric) and HCL (hydrochloric) should not be used in the wellbore after gravel packing. The HF and high HCL acid concentration must not be used in chrome pipes.

Resieved gravel-packing sand must be transported in special fifty-pound or one-hundred-pound sacks to minimize and avoid sand crushing. Gravel-packing sand must not be transported in sand transport trucks with high-pressure blowers.

Mishandling of gravel-packing sand and fracture-stimulation sand is a major issue. (I've seen sand blowers unloading sand into the sand master on a well location, crushing gravel dust over a mile long.) It is a poor job site without quality control, regard for the fracturing results, and/or workovers safety, sucking silica dust into their lungs!

There are several types of gravel-packing sand (props) that are used in sand control of today.

The gravel-packing sand may come from any part of the world without quality control

There are several regular types of silica gravel in the market:

> a. The Ottawa gravel-packing sand (named after a local area)—the sand appears to be round, with clarity and strength.
> b. The Brady sand—often used in fracturing, gravel packing, and industrial blasting.

Resieved sand should be used for gravel packing only!

The most popular sizes of reprocessed sand-gravel in the oil industry:

US Mesh Size	Sand-Gravel Size (in)
3–4	0.187 × 0.265
4–6	0.132 × 0.187
6–8	0.094 × 0.132
6–10	0.079 × 0.132
8–10	0.079 × 0.094
8–12	0.066 × 0.094
10–14	0.056 × 0.079
10–16	0.047 × 0.079
10–20	0.033 × 0.079
10–30	0.023 × 0.079
12–18	0.039 × 0.066
12–20	0.033 × 0.066
16–30	0.023 × 0.047
20–40	0.0165 × 0.033
30–40	0.0165 × 0.023
40–50	0.012 × 0.017
40–60	0.0098 × 0.017
40–70	0.0083 × 0.0165
40–100	0.0059 × 0.016

The most popular resieved sand used in gravel packing:

The 20–40 mesh (0.015″–0.030″)	permeability of 100 Darcy.
The 15–20 mesh (0.030″–0.040″)	permeability of 200 Darcy.
The 10–20 mesh (0.030″–0.060″)	permeability of 400 Darcy.
The 10–15 mesh (0.040″–0.060″)	permeability of 600 Darcy.

Other types and sizes of gravels may be used based on formation-sand analysis.

Reprocessed silica sand used in gravel packing is either angular or round shape.

Resieved and round gravel sands are mostly used in gravel-packing operation.

Do not apply fracture sand in gravel-packing operations. (Fracture sands are found to be unclean and contain crushed impurities!)

The reprocessed silica gravel-packing sand is subject to blast away and wash out with time from across the perforated holes because of production fluid movements and turbulence.

Resieved gravel-packing sand has higher porosity and permeability than most of the formation sands. Some reservoir formations produce large gravel sands (larger than 10/20 resieved gravel 0.030″–0.060″).

Some angular gravel-packing sand may have higher permeability and porosity than round sand gravels.

The following are resieved sand grades:

40/60	0.0098 × 0.0165
20/40	0.0165 × 0.0331
16/30	0.0471 × 0.0232

Note: Often glass-beads may be used in gravel-packing purposes. Glass beads are manufactured similar to gravel proppants used in the fracturing operations.

The compressive strength of glass beads is lower than gravel-packing sand and contains more fine particles. Glass beads are more rounded, and it has a smoother surface than gravel-packing sand.

Most glass beads are soluble in acid (so never flush with acid for any reason).

High-grade glass beads should only be used for gravel packing.

****

Fluid Filtration in Gravel Packing

As mentioned on previous pages, the most important steps in a well completion and gravel packing are as follows:

- ▶ Wellbore cleaning—techniques of removing mud filter and viscous-gelled material from the face of formation and out of perforated tunnels.

- ▶ Casing and tubing string cleaning—scale-removing techniques, pickling, and paraffin-removing techniques.

- ▶ Hot-oiling—it is not an effective process to remove the total paraffin out of tubing string. You will just pump the stuff back into the well again!

- ▶ Read the author's total paraffin-removing techniques later.

- ▶ Perforating techniques—high density, underbalanced perforating large holes.

- ▶ Equipment quality—screen design, gravel-packing sand quality and quantity.

- ▶ Use compatible clean, filtered completion fluid!

Most workover and/or drilling rig fluid tanks are not clean enough to store fluid for gravel packing and/or well-completion purpose.

Most workover fluid tanks contain dirty oil sludge, sand, dry dirt deposits, paraffin, drilling mud, cement, iron rust, and scale deposits inside the rig tank that must be removed or washed out with high-pressure steam jets. (Nasty housekeeping and careless operation!)

Vacuum trucks hauling water to the well site must be checked to make sure they are washed clean before transporting the gravel-packing fluid to the job site.

Produced water is found to be nasty and dirty with iron-sulfide, bugs, mud, sand, and bacteria. (They are practically hauling bacteria from one well to another well.)

You need to get good quality, clean and clear water to wash the wellbore in order to gravel pack and complete a well. (You need to train your people and the supervisors or send them home!) Some supervisors and contractors are not fit for sensitive oil-field work. They fake it, hoping to make it. (They cause irreversible damage and cost the oil industry a great deal.)

Some may say, "Why do we have to filter and pump clean fluid in a dirty well?" This thought process is very wrong; any microsolids you pump into the formation may plug off the tiny permeability and the porosity of the reservoir sand! Many oil-well reservoirs are lost because of bad completion fluids.

Note: the tiny reservoir pores are very sensitive to the nasty fluid that you plan to pump down the hole and will plug off the rock's permeability and porosity!

Filter units of various sizes and capacities are available in the gravel-packing industry. Filters with 1 micron absolute to 6 microns should be used to filter all the fluid during wellbore washing and gravel packing.

Treated bay water and gulf water must be filtered to 4 microns absolute for wellbore circulation and gravel-packing purposes. (Bay water contains a lot of suspended elements that will cause scale problems. Take a sample from the ship channel in Baytown, TX. or South Louisiana's bay and bayous and judge for yourself. You must use biocide to treat bay water.)

Using two filter units in tandem is recommended to make sure plenty of filtered water is available, without shutting down during gravel packing. Filter units may appear in horizontal or vertical vessels.

Changing out dirty filters during a workover is necessary and is recommended to avoid high-pumping pressure at the surface while gravel packing and to prevent dirty fluid from mixing with the grave-packing sand. (Haul off all your dirty filters after the project.)

As mentioned, before completion fluid is considered, check the most important components of the gravel-packing project.

Gravel-packing fluid must have the following characteristics:

— It can be used for well-control purpose (important).
— It is compatible with most reservoir formations (important).
— It is a nontoxic and nonhazardous fluid.
— It is free from solids and emulsions.
— It is a noncorrosive fluid.
— The brine fluid and/or diesel oil may be used as gravel-packing fluid.
— The produced production fluid (saltwater) and/or brine fluid is preferred for gravel packing because of its salt-content characteristics.
— Avoid using freshwater. Freshwater may cause clay swelling and dispersion of clay particles in the formation sand.
— Use inhibitor agents if necessary to prevent clay swelling. (Do not cut cost on this.)

The popular brine water for completion fluids are as follows:

KCL (potassium chloride)	8.4 to 10 pounds/gal
NaCL (sodium chloride)	8.30 to 10 pounds/gal
$CaCL_2$ (calcium chloride)	8.30 to 11.7 pounds/gal

Never use zinc chloride in gravel-packing operations.

There is always the possibility of premature screen plugging as the result of bad completion fluid.

****

The Gravel-Packing Units

The gravel-packing unit is a special and important equipment in water-pack gravel packing. The gravel-packing unit is designed to mix and hydraulically place the gravel-packing sand into the formation and around the screen and liner.

The gravel-packing unit consists of high-pressure mixing pots (drums or tubs), high pressure–rated plug valves, pressure gauges, and a fluid analyzer meter.

The mixing pots on the water-pack gravel-packing unit do not have high-speed blades or blender and will not crush any gravel-packing sand! In the water-based gravel-packing unit, the gravel-packing sand is packed in fifty-pound or one-hundred-pound bags. The bags are cut, and sand is manually poured into the pot before the top opening is closed shut. The lower-section plug valve is manually opened, causing the flow to stream to the top of the sand post. The differential pressure forces the gravel-packing sand to slurry through the tubing or annulus and to the perforations. There is no paddle-wheel blender or high-pressure hydraulic plunger pump to cause the gravel crushing in the water-pack gravel-packing units.

All the tools and equipment are built or mounted on a small trailer and transported to and from a well location. (Remove the fluid meter after each job before taking off.)

Calculated Required Volume of Gravel Sand

In water-pack gravel packing, fifty-pound or one-hundred-pound sacks of gravel-packing sand will be placed in the mixing pot and blended with filtered mixing completion fluid, such as brine water and/or field produced saltwater. (Do not strain your back when lifting the gravel-packing sand bags!)

Gravel-packing sand is normally packaged in one-hundred-pound or fifty-pound paper bags or plastic bags for easy lifting.

Hint: One hundred pounds of gravel-packing sand is equal to one cubic foot.
        Two thousand pounds is equal to one ton.

Experienced gravel-packing operators will have full control of sand volume, rate, velocity, pressure, and displacement volumes at all times. (I found these guys to be sharp with sufficient knowledge of gravel-packing calculation skills.)

Gravel-packing unit offers the best hydraulic method of placing an expected volume of gravel sand and filtered fluid with low-pump pressure and fluid velocity in Braden-head gravel-packing method.

Gravel-packing units are extensively used to spot or set a sand bridge in the deep or shallow wellbores without complications and with low cost.

Prepare to move a workover rig to conduct gravel packing!

I.   The Water-Based Gravel-Packing Methods in Oil-and-Gas Wells

Practical Application and Methodology

There are several methods of water-based gravel packing.

The past and the present gravel-packing methods are as follows:

A.   Cased-hole poor-boy gravel-packing method (reverse circulating pack)

B.   Cased-hole prepacking sand control using a crossover tool

C.   Cased-hole prepacking sand control using wash-down method

D.   Through-tubing gravel-packing method

E.   Oil-and-gas well open-hole gravel-packing method

There are basically three ways of installing or delivering a screen and liner in a cased perforated hole:

— reverse-circulating method (poor-boy gravel packing)

— crossover gravel-packing method

— wash-down method

The art of gravel packing is to control the formation sand and solids.

I-A. The Concept of Poor-Boy Gravel-Packing Sand Control

The poor-boy gravel packing was actually the first applied sand-control method in the oil fields and derived from the water well gravel-packing ideas.

As the name implies, the poor-boy gravel packing is a quick sand-packing procedure with low-cost gravel-packing method in water wells and/or shallow oil wells.

Poor-boy gravel packing is done in the shallow-depth wellbores with low pressure and low oil- and gas-producing wells with unconsolidated sand problems.

The poor-boy gravel-packing method is done quickly and is done based on a lower completion and operating cost.

The poor-boy gravel-packing procedure normally does not require extensive wellbore cleaning, prepacking, or any modern high-tech gravel-packing techniques.

The gravel-packing units and expensive slurry pack equipment will not be used in order to save time and money. (It is a simple and do-it-yourself project to cut cost!)

****

Presenting a Poor-Boy Gravel-Packing Procedure

The poor-boy gravel packing is basically a reverse-circulating gravel-packing method, and it is conducted in shallow-depth wells with low bottom-hole pressure, low production oil and/or gas wells with unconsolidated formation-sand problems.

These wells produce good oil-and-gas cut with low-water production and some gas production at a good profitable margin!

The fact about poor-boy gravel packing is proved to be trouble-free for a longer time than some slurry packings and/or the conventional sand-control methods.

Some of the poor-boy gravel packing may last fifteen years or longer in a wellbore if it is done correctly.

Prepare a workover rig and equipment to conduct a poor-boy gravel packing.

Safety Is Not an Accident!

The following operational safety tips will apply to all the gravel-packing completions and stimulation practices.

Move in and rig-up a workover rig and equipment to a well to conduct remedial sand control.

Over-the-Land or Shallow Bay-Water Operation

    A.   Verify the lease and the well name (do no pull the wrong well).

    B.   Verify and review the well information with the company man in charge.

    C.   Review the tubing string size, grade, and condition of the tubing rating.

    D.   Verify the derrick rating and drill-line condition before rigging up on a well.

    E.   Move in a workover rig to a dry and hard well location (if land work).

    F.   Check the well's location for hazards and overhead power lines. Move in a spot fluid pump and a clean rig tank.

G. The reverse unit package: fluid pump, tank, steel lines, swivel, stripper head. Steam clean the rig tank and fill with clean completion water.

H. Check and remove any fencing around the well prior to backing up to the well.

I. Use lock out and tag out if necessary. Install windsocks!

J. Check the location to ensure the ground is dry and hard (suitable to rig up). Use a steel mat to avoid sinking into the soft sand and clay on the well's location.

K. All the rig crew must wear their personal protective equipment (PPE), such as hard-toed boots, hard hats, eye-safety glasses, working gloves, clean and dry long-sleeved shirt, and pants (no rings and no necklaces or earrings!).

L. Check for toxic gas. Never assume a well or a location to be H_2S or CO_2 gas-free.

M. Start moving in and backing up the well. Use two spotters to guide the rig back to the wellhead. Do not run into the wellhead equipment and flow lines.

N. Rig up on the well. Hang blocks/pipe elevators straight at the center of the wellbore.

O. Do not rig up the rig within twenty feet of any energized power lines.

P. Tension-test guy-wire anchors (deadman), 23,000 lb. If using steel-based beams, the beam anchors must be certified to stay in compliance (use API guidelines).

Q. Failed safety anchors are subject to One Call or Dig-Test law.

R. No smoking within 150 feet from the wellhead.

S. You must check safety belts on each trip going up and down the derrick. Both hands must be free while climbing up the derrick!

T. Always maintain three-point contact while climbing up above the rig floor or coming down the derrick! (Never remove your safety belts while working above the floor.)

U. String out guy-lines to the safety anchors away from any power lines.

V. Spot reverse the unit (pump, tank, and power swivel).

W. Review the wellbore with a company man before you start working on the well.

X. Check and calibrate the weight indicator to avoid accidents (change if necessary).

Y. Check the drill line for flat spots and flags, and change if necessary.

Z. Use a body harness when working five feet above the rig floor!

AA. Conduct safety meeting before working on the well.

****

Wellbore Preparation before Remedial Gravel Packing

a. Open the well; read and record the shut-in tubing and shut-in casing pressure.

b. Bleed off the tubing and casing pressure slowly into the rig tank to 0 psi.

c. Circulate to kill well using filtered brine fluid.

d. Ensure that the casing valve/valves are fully open before nippling down or removing any wellhead equipment (a blind bull plug may be installed into the wellhead).

e. Never assume any oil or gas well is dead before tripping out and/or tripping into a well. Do not break the gas out of solution while tripping pipe.

f. Pipe movements will stimulate and energize the fluid, and it will cause the well to flow forcefully at once, causing pollution and possible accidents.

g. Check the well for flow. Circulate to kill the well first before removing the wellhead equipment.

h. Note: bleeding the gas pressure and bull-heading kill fluid through the tubing and down the casing string may not be good enough to kill some wells.

i. In case of a well flow, never use a plastic bucket to divert the fluid flow. Plastic buckets may cause fire and burn the rig (static electricity).

j. In case of sudden well flow, kill the rig first while the crew installs the safety valve on the tubing/drill string.

k. Pumping a few barrels of kill fluid down the tubing and/or casing may not be good enough to stop a well from flowing.

    Bullhead fluid is just a temporary kill. The well may flow on you again and again, and may cause unexpected accidents!

    Bullheading the kill fluid must be done properly by displacing the entire wellbore with kill fluid and/or circulating the influx out of the wellbore properly.

l. If the well is flowing oil, gas, and water, circulate to kill well using heavier fluid. (Wellbore fluid must stay static during the sand-control workover process.)

m. The proper way to kill a well is to circulate the well while the tubing is at the bottom of the well (provided there are no holes in the tubing string).

n. Rig up the floor and handling tools, and prepare to pull all the production equipment out of the wellbore (if any).

o. Unbeam the head, unseat the pump, and pull and lay down the polished rod. Pull out of the hole with a sucker rod string and fluid pump and/or artificial lift equipment (if any).

p. If a well is flowing heavy gas, do not pull rods. Shut down and kill the well properly. Ensure that the stripper rubber will fit and seal off tightly in case of emergency.

q. Check the well pressure and nipple down wellhead, and release the production equipment (packer, tubing anchor, seal assembly, etc.), if any.

r. Use short slips when needed to avoid unnecessary high-tension pulling into the used tubing string before releasing a tension set equipment! (This may avoid shearing the tools or ripping off the old casing!)

s. Many casings are damaged with holes and splits in the oil wells as a result of poor practice and lack of training in setting, releasing and nippling up/nippling down of the downhole tension tools. (You may cause the tubing strings to part. The well owner will never know what damages you have done to their well!)

t. Nipple up and test BOPs 200 psi low and 3,000 psi high (no shortcut on any BOP or safety equipment testing, regardless of type).

If the wellhead is below the ground level and/or inside a cellar, space out the BOP with an adapter spool to operate the BOP above the surface ground as a safety measure.

u. Never take shortcuts on safety. (Problems will catch up with you.)

v.   Keep a safety valve (TIW) on the rig floor in full open position with the wrench handle.

Teach your people how to use the safety valve during a flow emergency!
Never use a flow-line ball valve as a workover operation safety valve.
No L-shaped or angled-shape connections are allowed above any safety valves. It might cause fire and burn down the rig!

w.   POOH (pull out of the hole) with the tubing string and fill up the hole with kill fluid properly. The wellbore fluid must remain static (dead) during the entire workover operation.

x.   Rotate to release the mechanical packers and tubing anchor catcher (TAC) free. Always check the tool by slacking off.

y.   Rotate TAC two rounds every twenty stands while coming out the hole. (Avoid shearing the tool!)

z.   Make sure you understand how to set and release the production tools. Some mechanical tools are right-hand rotation to set and to release. Some tool are straight pull-release, and some mechanical tools are left-hand set and right-hand release. (Learn and work with the tools; do not fight with the tools.)
Tools are as good as the person who run and/or release them.

Note: All the mechanical tools are subject to re-set or hang up during a trip and may cause accidents. Do not allow the tool to rotate while tripping in or out of the hole.

- Move in and spot the fluid pump and clean the fluid rig tank.

- Fill up the rig tank with completion fluid and/or produced field saltwater or KCL water.

- Circulate wellbore to kill the well at any time with clean, filtered water, if necessary.

- Keep the hole full properly while tripping out of the hole.

- Note: Make certain that the tubing string and casing is free of paraffin, formation scales, salt, and asphaltic elements. Unclean casing and tubing will release paraffin, asphaltic elements, and scale deposits, and they will mix with the gravel-packing sand!

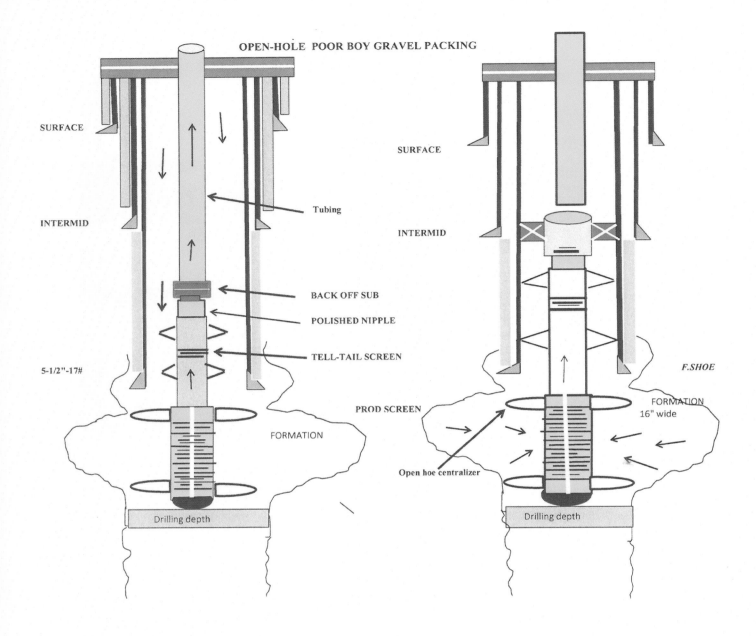

OPEN-HOLE POOR BOY GRAVEL PACKING

The poor-boy gravel packing is done in six simple steps:

1. Run and clean up the wellbore using a rock bit and scraper or a mule shoe joint.

2. Run and set a cast-iron bridge plug or hard cement base to establish a foot base for the screen and liner to set on.

3. Run and set the gravel-packing screen and liner at the bottom, on the base plug.

4. Pour and displace resieved gravel-packing sand around the screen and liner.

5. Release from the screen and liner assembly.

6. Run the production equipment and place the well on production.

Isolation packer is an advantage option in poor-boy gravel packing.

Application of a Mule Shoe or Rock Bit before Gravel Packing

The following are applications of mule shoe in the well washing and gravel packing:

- The mule shoe is a joint tubing or piece of open-ended pipe that is cut in an angle at one end, shaping like a mule's shoe (slant-cut pipe).

- Mule shoe joints have extensive applications in washing and circulating sand and solids out of a wellbore in oil wells (an inexpensive tool to use).

- The application of a mule shoe is a low-cost practice to clean up a wellbore. (Do not attempt to misuse, drill, or rotate on hard spots with a mule shoe joint.)

  A notched tubing collar may also be used in borehole cleaning.

- The mule shoe is an open-ended pipe, and it is not designed to drill or to be used as a mechanical or hydraulic jarring tool. Misusing and the wrong application of a mule shoe is an unprofessional practice.

- Forcing the mule shoe through perforations or tight spots may cause the shoe to sidetrack. Misapplication of the mule shoe will cause the pipe to become stuck and create fishing work. (You may find your mule shoe outside of the casing.)

- Trip in the hole with thirty-one feet of slick mule shoe joint on the work string. Test the tubing string in the hole. (You might not be able to wash solids with a hole in your tubing string.)

- Wash and circulate the wellbore to the required plug back depth below the open perforations. (Below the perforations must be cleaned up to set a bridge plug or to set a hard bottom as footing.)

- POOH with the tubing and mule shoe joint while filling up the hole with water (make certain that the hole is kept full with fluid at all times, if possible).

- If running a bit and a casing scraper, do not rotate or drill with the casing scraper. (Do not force the scraper through tight spots going down the hole into the perforation!) Always space out the bit and scraper. Do not drill with the scraper in the hole. Drilling with the scraper will damage the casing string.

A full-sized tri-cone rock bit is a safe tool to run and to drift a casing string. Rock bits have the ability to pass through the doglegs and casing curvatures. (It is not a stiff connection.)
A gauge ring and junk basket is the best method of drifting a casing string before gravel packing.

Run and set an isolation bridge plug (foot base).

- A hard foot base is necessary to set the screen and liner in the casing string below the open perforations.

- Rig up the electric wireline unit.

- Nipple up and install the casing lubricator or pack-off (as required).

- Trip in the hole and drift casing and open perforation with a gauge ring and junk basket.
  (Running a gauge ring and junk basket is a good practice. The gauge ring will drift the casing and perforations and will provide information for the screen and liner to pass through.)

- Clear and drift the casing string to twenty feet below the open perforations.

- POOH and lay down the gauge ring and junk basket.

- Trip in the hole with a cast-iron bridge plug (CIBP) on the electric line.

- Correlate and set the cast-iron bridge plug eight to ten feet below the bottom set of open perforations. (Keep a record of the bridge plug's landing depth for reference.)

- A cast-iron bridge plug may be run and set on a tubing string at any desired depth if necessary.

- POOH and rig down the wireline tools and equipment.

- Make a trip with the mule shoe joint to the top of the cast-iron bridge plug. Circulate the wellbore clean. (Use the bridge plug depth as reference, and mark the work string at the rig floor.)

- If the wellbore stays clean to the top of the cast-iron bridge plug (CIBP), POOH with the tubing string while keeping the hole full.

- Rig up the gravel-packing tools and equipment while tripping out of the well.

****

Presenting a Poor-Boy Gravel-Packing Sand Control

Cost cutting is an obstacle of not doing the job right. It might cost you later.

The poor-boy gravel packing is a reverse circulating gravel-packing method.

- Ensure a good-condition, casing and tubing string without leaks or holes.

- Ensure a clean wellbore to the top of the cast-iron bridge plug!

- Continue pumping filtered water in the well at a slow rate while preparing the screen and liner.

- Unload the screen and liner joints and set above the ground. The screen and liner will arrive on location in wood strip boxes and wrapped with plastic cover sheet to protect and avoid damages. Exercise caution when loading and unloading the screen and liner joints.

- Remove the wooden straps and unwrap the plastic; inspect and measure to make certain that the screen and liner are in good condition and without damages.

- Unwrap and check the screen gaps with a filler gauge. (If no filler gauge is available, pick up one spoonful of selected gravel-packing sand and put it on the screen gaps. Force sand gravel with your fingers into and between the wraps. The selected reseived sand should not go through the screen openings.)

- Measure the screen and liner, and keep an accurate tally of anything that goes into the well. (Keep a record of all the tubing joints on the location to avoid mistakes of miscounting.)

- Know how many joints or how many feet of tubing would take to land the screen and liner at the correct depth before starting into the well.

- Install special lift nubbins (lift sub) and pipe clamps on the blank space of the screen section. Pick up the screen above the rig floor. Mark and number the screen/screens and liner joints in an orderly manner when picking up and running into the well.

- Make sure the screen section is overlapped at least five feet below and five feet above the open perforations. (This will get you off the rope when you reach the bottom.)

- Avoid any damages to the screen and liner assembly while unloading, lifting, and making up. (Make the screen and liner by hand using friction wrenches only.)

- Drift and clear anything that goes into the well using a Teflon rabbit.

The Poor-Boy Gravel-Packing

The Poor-Boy gravel packing screen assembly consists of:

    a)   Blind bull plug
    b)   Sand screen section
    c)   Blank tubing joint
    d)   The tell-tail screen joint
    e)   The polished nipple
    f)   The releasing back-off sub
    g)   Work string tubing

Pick-up and start going into the hole with the screen and liner.

    1.   A blind bull plug

A bull plug with welded flat circle plate or bladed bottom is preferred.

    2.   Screen joints

Make sure the screen section is long enough to cover the entire open perforations plus five feet below and five feet above the open perforation.

3. Blank liners

    You may use two joints, three joints, or as many joints as needed above the screen section.

4. Telltale screen joint (your important downhole camera)
    The telltale screen joint is made on a standard blank pipe of thirty-two feet long with two feet of screen section located twenty feet from the top.
    Telltale is an important joint to tell you the height of the gravel-packing sand in the annulus, above the screen section.

5. The polished nipple
    The polished back-off sub (polished nipple) is an important special sub screwed on top of a telltale screen joint. It is designed to run the screen and liner in the well.
    The polished nipple has a pin on one end and female box-thread on the other end.

The tool has an acme left-hand coarse thread box on the top (looking up) and regular or 8rd thread pin at the bottom to fit the collar on the screen and liner (the telltale joint).

The polished back-off sub will remain in the hole on the top of liner assembly of the screen and liner after the work string is released, and the gravel pack is completed!

The polished nipples are either polished or coated with epoxy paint to last longer. Use two feet or longer nipple as a fishing neck above the screen and liner. The polished nipple is located on top of the telltale joint of the screen and liner.

Polished nipples are machined in various sizes and lengths of thirty-six inches for all practical purposes. (This will help to fish the screen and liner from the outside instead of screwing into the nipple several years later. After several years, the inside threads [acme] may be washed or corroded away and will be difficult to screw into the nipple to retrieve the old screen and liner out of the well.)

Polished nipples will also be used as a neck to run the O-ring overshot pack off to isolate the screen and liner, if it is needed.

6. Releasing crossover sub (the releasing tool)

The releasing tool is basically a short crossover nipple (crossover swage). The crossover sub has an acme thread on the pin-end at the bottom to fit the polished nipple, and an 8rd thread in the boxed end at the top to match the work string pipe (various sizes are available for the job).

The releasing tool is a left-hand makeup and a right-hand releasing tool.

The releasing back-off sub will be screwed at the bottom of the work string and onto the polished nipple.

(Turning the sub to the left will tighten it, and turning it eight to twelve rounds to the right will back off or release it from the polished nipple on the screen/liner.)
It's a very simple tool for the trained and knowledgeable operator.
The releasing back-off tool will be screwed at the bottom of a crossover tool and/or work string and onto the polished nipple to run the screen and liner into the well.

- Poor-boy gravel packing may be done with or without a crossover tool.

- Fully open the rams in the BOP stack (make sure the rams are the correct size).

- If you know the static fluid level, you may calculate the bottom-hole pressure.

- Continue running water into the annulus slowly at all times.

- The wellbore must stay static or on a slight vacuum while running the screen and liner in the well.

- You do not want the well to flow back while going into the hole. Fluid flowing back means that possible formation sand is caving into the wellbore with the fluid flow (the screen and liner might not get down). There is a significant difference between a well flowing and a pipe displacement. (Learn the concept to save the crew from the well kicks!)

- Remove any and all the plastic wraps outside of screen sections. Pick up and lower the screen into the borehole using lift subs and special pipe clamps.

- Install special lift nubbins and pipe clamp on the screen/screens to lift the screen and liner using a flat-based plate. (Do not use a pipe clamp on the screen section!)

- Pipe clamps used on gravel packing are different from the pipe clamps used in fishing operations. Pipe clamps used in gravel packing consist of two pieces of flat steel bars that bend, and they are forged in a circular shape to fit and lift small pipes with light weight.
Holes and bolts are used to keep the clamp tight around the pipe when used as lifting elevator.

Pipe clamps are forged to fit 1″ through 2-⅞″.

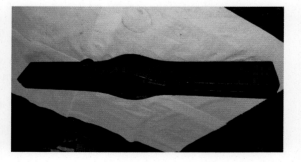

- Remove thread protectors on the rig floor. Do not drag the screen and liner on the ground!

- Pick up the liner joints with subs and lift above the rig floor (one joint at a time and in the running order).

  Lower the screen into the hole slowly while checking the screen. (Keep the screen clean and free from mud and the pipe dope on your dirty gloves.)

- Unwrap and remove any and all plastic wraps while lowering the screen into the hole. Apply light pipe dope on the pin-end threads only.

- Make up the liner joints onto the screen section using a friction pipe wrench to avoid damaging the pipe coating and avoid galling. Lower the screen/liner slowly into the well using flat-based plates, lift nubs, and pipe clamps.

- Pick up and check the telltale joint with the polished nipple on top of the liner assembly. Make up the telltale joint onto the liner assembly.

- Install the polished nipple onto the telltale joint tightly. (The polished nipple will stay in the well after the gravel packing.)

- Make up the right-hand-release back-off sub on a joint of tubing, and screw the back-off sub onto the polished nipple using a hand pipe wrench only (make up, bump, back off, and make up, snug tight again). An experienced gravel-packing operator should know this important step!

- Check the final makeup snug bump with a hand pipe wrench. Lower the screen and liner into the well. Do not make up the screen and liner with power tongs!

- An experienced gravel-packing man should know and feel the tightness on the releasing sub. (Threads must be snug but not too tight to release.)

- Never make up the back-off sub into the polished nipple too tight! It will be difficult to back off from the screen and liner. (Pay attention!)

  The back-off sub has acme threads; it is left-hand set and right-hand turn to release.
  When making (turning) the pipe to the right, make sure you do not release the back-off sub from the screen and liner. Use the back-ups to avoid the pipe from right-hand rotation.

- Lower the screen and liner into the well with the screen and liner, and continue in the hole on the work string tubing, going slowly down the hole.

- Make the work string tubing joints by hand first, then apply hydraulic tongs to the proper makeup. (Make sure the tubing is in good condition, without any holes!)

- Tag the bottom on the cast-iron bridge plug lightly (foot base) and pick up six inches off the bottom. Check the pipe measurements to make sure the screen landing is accurate and the screen section has landed across the entire open perforations.

- Rig up the pump, tanks, and break-fluid circulation by reverse circulating (pumping fluid down the casing annulus and out of the screen and tubing string).

The annulus is the space between the inside of casing and the outside of screen and liner or tubing string.

When pumping down the annulus, the maximum applied pressure will be on the casing.
When pumping down tubing string, the maximum applied pressure will be on the tubing string.

  ► Pumping down the casing is referred to as short way or reverse circulation.

  ► Pumping down the tubing and screen and liner is referred to as long way or conventional circulation.

What else do you want to know?

- When breaking the fluid circulation, water will go through the annulus, between the casing and the tubing, and will come out through the telltale joint or screen section, and it will circulate up the tubing string to the surface.

- Calculating the fluid volume and gravel-packing sand is necessary to bring the gravel sand from the top of the cast-iron bridge plug and up the annulus to the telltale joint (twenty feet below the top of blank liner on the telltale joint).

- Break circulation and establish pumping rate down the annulus, up through the screen/liner and onto the rig tank. (Keep the hole full with water at all times, if possible.)

- Note:

    The basic calculation and knowledge of fluid measurement and sand-volume requirement are necessary to conduct a gravel-packing operation.

    Pipe capacity, displacement, volume, and heights between two pipe strings must be known as accurately as possible to avoid complicated problems.
    See ready-made calculated values and formulas for reference.

- Rig up the fluid pump and tank. Break circulation to establish base fluid. Start pouring, mixing, and displacing the required calculated volume of gravel-packing sand slowly down the casing annulus.

- Mix the gravel-packing sand slowly with a good continuous stream of clean water. (Watch the fluid return out of the tubing string while pouring and displacing sand grains with water down the casing annulus.) You may use large funnels for pouring sand for consistency purposes.

- Never stop the mixing water while pouring sand down the annulus (it may cause a fishing job)!

- Continue mixing and displacing gravel-packing sand slowly down the annulus of the casing and screen and liner (or tubing string).

- When pumping gravel down the annulus around the screen and liner, some of the gravel-packing sand may actually be forced out through the perforation tunnels and into the formation because of hydrostatic head of water and sand. (Have plenty of sand and water on location at all times.)

- Never run out of water when pouring sand. It will get the tubing stuck!

- If the well goes on hard vacuum, stop dumping sand and continue with water down the annulus until the well stops sucking fluid (avoid sand bridge).

- Once all the required calculated volume of gravel-packing sand is poured down the casing annulus, close the BOP shut while pumping water, and monitor fluid displacement and return out of the tubing string.

    Flush and displace gravel-packing sand with clean water at a slow pumping rate down the annulus while watching the annulus pressure rise.

- Continue displacing gravel-packing sand down the casing string until the gravel-packing sand reaches the bottom of the screen in the annulus to the telltale screen section.

- Once the annulus between the casing and screen and liner is filled with gravel-packing sand, the pressure in the annulus will increase, causing the fluid-circulation returns to slow down at surface.

- The increase in the annulus pressure will indicate a good sand pack from the end of screen and liner to the midpoint of the telltale joint. (The telltale screen is only two feet long and designed to tell you where the top of the sand gravel may be.)

- Stress the gravel-packing sand with 500 psi once. Shut down the pump and wait for one hour. Continue pumping water down the annulus at a low rate, and re-stress sand pack to 700 psi or as necessary. (You will see small fluid circulation at the surface.)

- If the formation took some of the sand or is taking sand, top off the sand pack to the telltale again, if necessary. (Avoid dumping too much sand at one time during sand packing. Adding more sand is basically a hard judgment call.)

- Never dump too much sand at one time. This will bring the sand to the top of the back-off sub and create problems when backing off from the screen and liner.

- If gravel packing is satisfactory, make a final pressure pack and simply pick up on the work string slowly eight hundred pounds to one thousand pounds upstream (eight hundred pounds above the weight of tubing string). Rotate the tubing eight to twelve right-hand turns at the tool to release, and back off from the screen and liner assembly. (This will indicate good sand packing).

- Be sure not to drag the screen and liner up the hole when backing off from the screen and liner. (The screen should not move unless there is a void and/or not enough sand above the screen section.)

  If the screen moved up the hole, it will be difficult to push the screen and liner back to the correct landing position.

- Pick up on the work string six inches off the polished nipple after releasing from the screen and liner, and reverse circulate the wellbore immediately (down the annulus and out of the tubing string).

- Circulate two tubing volumes clean while checking for any sand return. (Use a pair of socks on the return line to catch samples. There must not be any sand grains in the sample catcher.)

- POOH with the tubing string and lay down the right-hand releasing tool.

- The polished nipple will stay in the hole on top of the screen/liner until it is time to fish the screen out of the wellbore.

  ▶ Note: a gravel-packing unit and a crossover tool can be used to facilitate the poor-boy gravel-packing method safely, quickly, and more efficiently.

- Isolation packers, such as a model-L packer or others, are additional cost-related options in poor-boy gravel packing.

- Trip in the hole with the production equipment and put the well on production.

- The polished nipple on the screen and liner will stick up in the center of the production casing with lots of gravel-packing sand below it.

The following are advantages and disadvantages of poor-boy gravel packing:

▶ A sand bridge may occur when pouring sand down the annulus. This will happen if you are pouring the sand too fast or the well may go on a vacuum (you will get the tubing stuck).

▶ Use a crossover tool to reduce gravel-packing bridge in deeper wells.

▶ It is difficult to tag bottom and check inside of the screen and liner without a packer.

▶ It is difficult to get down with a wireline. (You may use a shear pin overshot to get through.)

▶ In the low bottom-hole pressure, gravel-packed wellbore, the poor-boy sand pack will last many years without problems.

▶ If the poor-boy sand-packing job is done correctly, the screen and liner may stay in the well for fifteen years without complication problems.

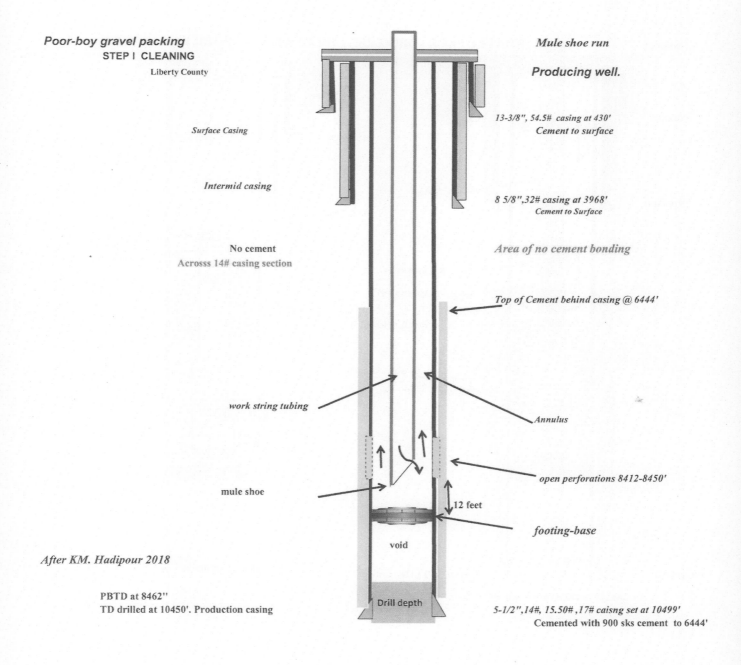

Poor-boy gravel packing
STEP I CLEANING
Liberty County

Mule shoe run

Producing well.

Surface Casing

13-3/8", 54.5# casing at 430'
Cement to surface

Intermid casing

8 5/8",32# casing at 3968'
Cement to Surface

No cement
Acrosss 14# casing section

Area of no cement bonding

Top of Cement behind casing @ 6444'

work string tubing

Annulus

open perforations 8412-8450'

mule shoe

12 feet

footing-base

void

After KM. Hadipour 2018

PBTD at 8462"
TD drilled at 10450'. Production casing

Drill depth

5-1/2",14#, 15.50# ,17# caisng set at 10499'
Cemented with 900 sks cement to 6444'

73

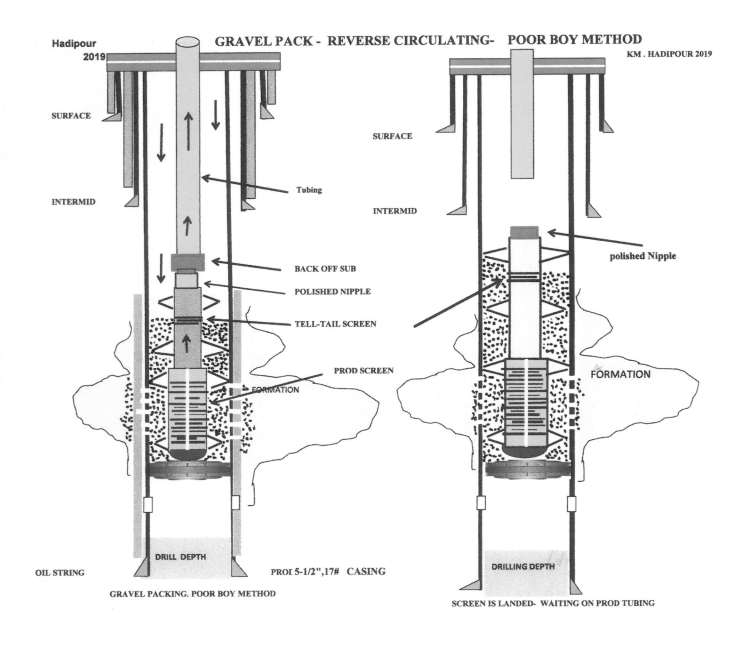

GRAVEL PACK - REVERSE CIRCULATING- POOR BOY METHOD

Hadipour 2019

KM . HADIPOUR 2019

SURFACE

INTERMID

SURFACE

INTERMID

Tubing

polished Nipple

BACK OFF SUB

POLISHED NIPPLE

TELL-TAIL SCREEN

PROD SCREEN

FORMATION

FORMATION

DRILL DEPTH

DRILLING DEPTH

OIL STRING

PROI 5-1/2",17# CASING

GRAVEL PACKING. POOR BOY METHOD

SCREEN IS LANDED- WAITING ON PROD TUBING

Basic Important Formulas

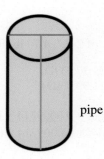

pipe

- hydrostatic pressure = 0.0519 × fluid height × fluid density
- area of tubing (circle) (pipe inside diameter)^2 × 0.7854
- area of tubing (circle) (pipe radius)^2 × 3.1416 [π]
- pressure = force divided by cross-section area
- hydrostatic head = the weight of column of fluid at rest
- R = radius of the circle
- D = diameter of the pipe
- Hint: 1 gallon of freshwater = 1 pound = 231 cubic inches
- 100 pounds of sand = 1 cubic foot
- 2,000 pounds = 1 ton
- 1 gallon = 3.785 liters
- All the oil- and gas-production reservoirs have pressure.
- Some reservoirs will have a significant high pressure, and some wellbores will have lower pressure.
- If the reservoir pressure is higher than the hydrostatic head fluid in the tubing/casing, the well will flow.
- If the reservoir pressure is equal with the hydrostatic head of fluid in the tubing/casing, the well fluid will stay static at different heights below the surface.
- When you push the pipe into the fluid in the wellbore, you may raise the fluid level and the hydrostatic head (fluid may spill or runover).
- Gas trapped in the solution is dangerous and may cause kicks.
- Low fluid level in a well is indicative of low reservoir pressure, a depleted reservoir, or a sanded-up wellbore (perforations are covered up).

Important Terms Used in the Oil Field

Capacity and Displacement

Capacity is "the ability to hold or ability to contain within." *Capacity* is "the volume of a certain tube to hold fluid [tubing, casing, and stock tank capacity]."

Displacement is basically "replacement measurement." (When lowering the tubing string into a well full of fluid, the excess water that spills out is the replacement or tubing displacement.)

There is a significant difference between pipe displacement while going into a well and a natural well-flow pattern. (It is important to understand and remember that to avoid blowout accidents.)

Always check the well for flow while tripping a pipe in a well (early detection and a quick reaction).

Basic Important Calculations

Tubing Capacity

Example: capacity of 2-⅜", 4.70 lb./ft. tubing or pipe

(2.375"OD)
(1.995" ID)

Barrel per linear foot = 0.0009714 × D²   D = the inside diameter of tube
Barrel per lineal foot = 0.0009714 × (1.995)² = 0.003865

Barrel per lineal foot = 0.003865 barrels = capacity of 2-⅜", 4.70# tubing

Capacity per 1,000′ = 0.9714 × (1.995)² = 3.87 barrels
or 1,000′ × 0.003865 = 3.87 barrels

Linear feet per one barrel = $\dfrac{(1029.4)}{D^2 = (1.995)^2}$ = 258.60 feet

or 1 bbl. divided by 0.003865 = 258.60 feet

The capacity of casing or tubing (barrels per foot) = 0.0009714 × (inside diameter of pipe)²

Liter per meter = $\dfrac{D^2}{1273}$ Meter per liter = $\dfrac{1273}{D^2}$

Learn MKS system; it is more accurate than CGS system!

pipe   h

The Tubing String Capacity

Tubing size	Weight (lb./ft.)	Inside diameter	Capacity (bbl./ft.)	Cu. Ft. per Lin. Ft.	
1.900" (1.610)	2.90	1.61	0.0025	0.0149	70.70′
2.063" (1.751)	3.25	1.751	0.0030	0.0167	59.80′
2-⅜" (2.375)	4.70	1.995	0.00387	0.02171	46.10′
2-⅞" (2.875)	6.50	2.441	0.00578	0.0325	30.77′
2-⅞" (2.875)	8.60	2.259	0.00520	0.02783	35.93′
3-½" (3.500)	9.30	2.992	0.00870	0.04884	20.48′
3-½" (3.500)	10.20	2.992	0.00830	0.0466	21.47′
4"	9.50	3.43	0.0123	0.0532	10.13′
4"	11.0	3.35	0.0118	0.0511	10.11′
4"	11.60	3.31	0.0114	0.0510	10.14′
4-½" (4.500)	12.75	3.598	0.0152	0.0854	11.70′

Cubic foot per linear foot is obtained from:

Example: what is the cubic foot per linear foot in 2-3/8" tubing? 0.00387 divided by 0.1781 = .0217 or 0.00387 divided by 5.6 = .0217

1 cubic foot in 2-3/8" tubing is: 1 divided by 0.0217 = 46.10

The Casing Capacities

4-½"	9.50#	4.09	0.0162	0.0912	10.96'
4-½"	10.50#	4.052	0.0159	0.0895	11.17'
4-½"	11.60#	4.000	0.0155	0.0872	11.46'
4-½"	12.60#	3.958	0.0152	0.0835	11.83'
4-½"	13.50#	3.920	0.0149	0.0832	11.93'
5"	15#	4.408	0.0189	0.1059	9.44'
5"	18#	4.276	0.0178	0.0997	10.03'
5-½"	14#	5.012	0.0244	0.1387	7.21'
5-½"	15.50#	4.960	0.0238	0.i336	7.48'
5-½"	17#	4.892	0.0232	0.1305	7.66'
5-½"	20#	4.778	0.0222	0.1245	8.03'
7"	17#	6.538	0.0415	0.2332	4.29'
7"	20#	6.456	0.0405	0.2273	4.40'
7"	23#	6.366	0.0393	0.2210	4.52'
7"	26#	6.276	0.0383	0.2148	4.66'
7"	32#	6.094	0.0361	0.2025	4.94'
7-⅝"	26#	6.844	0.0472	0.2646	3.88'
7-⅝"	29.70#	6.873	0.0458	0.2582	3.86'
7-⅝"	33.70#	6.640	0.0445	0.2496	4.01'
8-⅝"	24#	8.097	0.0637	0.3575	2.80'
8-⅝"	28#	8.017	0.0624	0.3505	2.85'
8-⅝"	32#	7.921	0.0609	0.3422	2.92'
8-⅝"	36#	7.825	0 .0595	0.3340	2.99'
9-⅝"	29.30#	9.063	0.0797	0.4479	2.23'
9-⅝"	32.30#	9.001	0.0787	0.4418	2.26'
9-⅝"	36#	8.921	0.0773	0.4340	2.31'
9-⅝"	38#	8.885	0.0766	0.4305	2.32'
9-⅝"	40#	8.835	0.0758	0.457	2.35'
9-⅝"	43.50#	8.755	0.0745	0.4180	2.39'
9-⅝"	47#	8.681	0.0732	0.4110	2.42'

How to use the above ready data:

Example 1
What is the capacity of 2,300 feet of 5½", 17# casing?

$$2300' \times 0.0232 = 53.56 \text{ barrels}$$

Example II
What is the capacity of 2,300 feet of 2⅞", 6.50# tubing?

$$2300' ' \times 0.00578 = 13.29 \text{ barrels}$$

Example III
How much 20/40 sand is needed to fill 100' in the 5-½", 15.50# casing?
0.1336 cu. ft. / liner ft. × 100 = 13.36 cubic foot or 13.36 sacks (100# sacks)

One cubic foot of 20/40 sand is equal to 100 pounds of sand.
100 pounds of 20/40 gravel-packing sand is equal to one cubic foot.

What is the hydrostatic pressure of 8,000 feet of 11 pound/gallon of calcium chloride water in 7″, 23# casing?

HP = 0.052 × 8,000 × 11 = 4,576 psi

One gallon of freshwater is equal to one pound and equal to 231 cubic feet.

What is the hydrostatic pressure of 8,000′ of 11 pound/gallon of calcium chloride water in 2″, 4.6# tubing?

HP = 0.052 × 8,000′ × 11= 4,576 psi

Typical Oil-and-Gas Wellbore Schematic

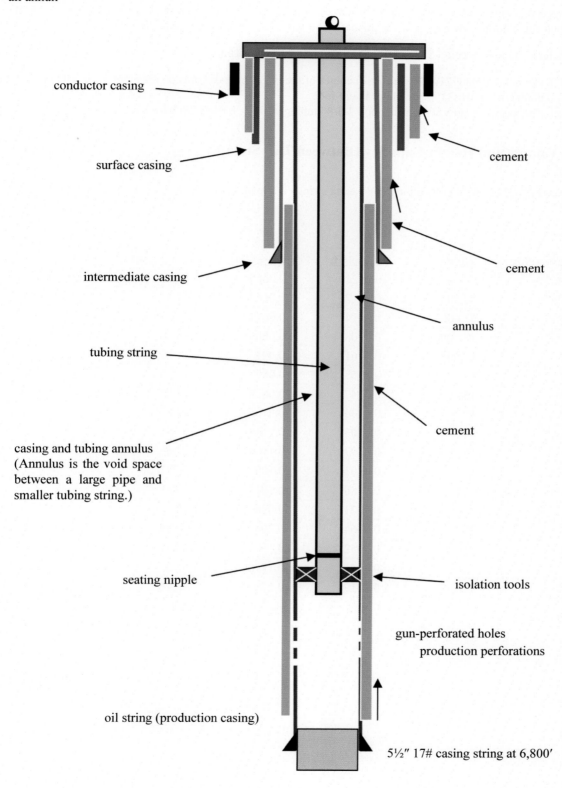

all annuli

conductor casing

surface casing

cement

intermediate casing

cement

annulus

tubing string

cement

casing and tubing annulus
(Annulus is the void space
between a large pipe and
smaller tubing string.)

seating nipple

isolation tools

gun-perforated holes
production perforations

oil string (production casing)

5½″ 17# casing string at 6,800′

K. M. Hadipour

Volume and Height between the Casing String and Tubing String

The volume and height are basically the capacity of annulus between a larger pipe and a smaller tubing or a volume between a pipe and an open hole.

The volume between the outside of a tubing and the inside of a casing is extremely important in gravel packing and the wellbore-circulation calculations (the casing inside diameter and the screen and liner outside diameter).

Example: the annular volume between 5.5″ (15.50#) casing and 2⅜″ tubing string.

(D=5.5″ d=2.375″) (D^2-d^2) × 0.0009714 (D = inside diameter of the larger pipe d = outside diameter of the smaller pipe)
$(24.50 - 5.641) \times 0.0009714 = 0.0183196$ barrel per linear foot
The annular capacity between 5.5″ (15.50#) casing and 2-3/8″ tubing string = 0.0183196 barrels.

The Annulus Volume and Height Capacity (Between Two Strings)

The volume between 2-$\frac{1}{16}$″ (2.0625″) OD tubing and casing strings is shown below:

Casing

Casing Size	Wt. per Ft.	Barrel/Ft.	Cu. Ft./Ft.	Lin. Ft. per Cu. Ft.
4-½″	9.50#	0.0121	0.0680	14.70′
4-½″	10.50#	0.0112	0.0663	15.07′
4-½″	11.60#	0.0114	0.0641	15.61′
4-½″	13.50#	0.0108	0.0606	16.50′
5″	11.50#	0.0161	0.091	11.09′
5″	13#	0.0155	0.087	11.50′
5″	15#	0.0147	0.0828	12.08′
5″	18#	0.0136	0.0765	13.07′
5-½″	14#	0.0203	0.1138	8.79′
5-½″	15.50#	0.0197	0.1104	9.06′
5-½″	17#	0.0191	0.108	9.32′
5-½″	20#	0.0180	0.1013	9.87′
5-½″	23#	0.0171	0.096	10.45′
7″	20#	0.0364	0.2041	4.899′
7″	23#	0.0352	0.199	5.055′
7″	26#	0.0341	0.1916	5.22′
7″	32#	0.0319	0.1793	5.576′

The Annular Capacity (between Two Strings)

The *annular capacity* is defined as "the volume between a casing string and tubing and/or the volume between an open hole and the drill string."

The volume between 2-⅜″ (2.375) OD tubing and casing strings is listed below:

Size /Wt. per Ft.	ID-Drift	Bbl./Ft.	Cu. Ft./Lin. Ft.	Lin. Ft. /Cu. Ft.
4-½″ (9.50#)	4.090″	0.0108	0.0605	16.54′
4-½″ (10.50#)	4.052″	0.0105	0.0588	17.1′
4-½″ (11.60#)	4.000″	0.0101	0.0565	17.70′
4-½″ (13.50#)	3.920″	0.0094	0.0530	18.85′
4-1/2″ (15.10)	3.826″	0.0087	0.0490	20.38′
5″ (11.50#)	4.560″	0.0147	0.0826	12.10′
5″ (13#)	4.369″	0.0141	0.0794	12.59′
5″ (15#)	4.404″	0.0134	0.0752	13.30′
5″ (18#)	4.277″	0.0123	0.0690	14.51′
5-½″ (13#)	5.044″	0.0192	0.1080	9.26′
5-½″ (14#)	5.013″	0.0189	0.1061	9.412′
5-½″ (15.50#)	4.950″	0.0183	0.1029	9.72′
5-½″ (17#)	4.893″	0.0178	0.0998	10.024′
5-½″ (20#)	4.777″	0.0167	0.0937	10.667′
5-½″ (23#)	4.670″	0.0157	0.0882	11.34′
7″ (17#)	6.536″	0.0360	0.2027	4.94′
7″ (20#)	6.455″	0.0343	0.1925	5.09′
7″ (23#)	6.366″	0.0339	0.1903	5.26′
7″ (26#)	6.276″	0.0326	0.1840	5.44′
7″ (32#)	6.095″	0.0306	0.1717	5.80′

Example: 70 barrels of water will fill up how many feet in the annulus between 2-⅜″, 4.70# tubing and 5-½″, 17# casing string?
 70 barrels divided by 0.0178 = 3,933 feet

How many barrels of fluid are needed to fill up 7,000′ of annular space between 2-⅜″ tubing and 5-½″, 17# casing?

0.0178 × 7,000′ = 124.6 barrels

The volume between 2-⅞″ (2.875″ OD) tubing and casing strings is listed below:

Size	Wt./ Ft.	Bbl./Ft.	Cu. Ft./Lin. Ft.	Lin. Ft./Cu. Ft.
4-½″	9.50#	0.0082	0.0462	21.67′
4-½″	10.50#	0.0079	0.445	22.49′
4-½″	11.60#	0.0075	0.0422	23.71′
4-½″	13.50#	0.0069	0.0387	25.82′
5″	11.50#	0.0122	0.0683	14.63′
5″	13#	0.0116	0.0651	15.37′
5″	15#	0.0108	0.0609	16.42′
5″	18#	0.0087	0.0546	18.30′
5-½″	13#	0.0167	0.0937	10.671′
5-½″	14#	0.0164	0.0651	15.37′
5-½″	15.50#	0.0158	0.0866	11.29′
5-½″	17#	0.0152	0.0854	11.703′
7″	17#	0.0335	0.1881	5.32′
7″	20#	0.0325	0.1545	6.47′
7″	23#	0.0313	0.1760	5.68′
7″	26#	0.0302	0.1697	5.89′
7″	32#	0.0280	0.1575	6.35′
7-⅝″	24#	0.0399	0.2241	4.46′
7-⅝″	26.40#	0.0391	0.220	4.529′
7-⅝″	29.70#	0.0379	0.2127	4.70′
7-⅝″	33.70	0.0364	0.2045	4.90′
7-⅝″	39#	0.0346	0.1943	5.147′

How many barrels of fluid are needed to fill up 9,000′ of the annular space between 2⅞″ tubing and 5½″, 17# casing?

9,000′ × 0.01512 = 136.8 bbls. (from table)

Volume and height between 3½″ (3.50″) OD tubing and casing strings are listed below:

Size	Wt./Ft.	Bbl./Ft.	Cu. Ft./Lin. Ft.	Lin. Ft./Cu. Ft.
7″	17#	0.0296	0.1663	6.02′
7″	20#	0.0286	0.1605	6.23′
7″	23#	0.0275	0.1542	6.48′
7″	26#	0.0264	0.1480	6.76′
7″	29#	0.0252	0.1418	7.05′
7″	32#	0.0242	0.1357	7.38′
7-⅝″	24#	0.0360	0.2023	4.94′
7-⅝″	26#	0.0353	0.1981	5.49′
7-⅝″	29.70#	0.0340	0.1921	5.24′
7-⅝″	33.70#	0.0326	0.1828	5.47′

Two hundred barrels of water will fill up how many feet of annulus space between 3-½″ tubing and 7-⅝″, 26# casing string?

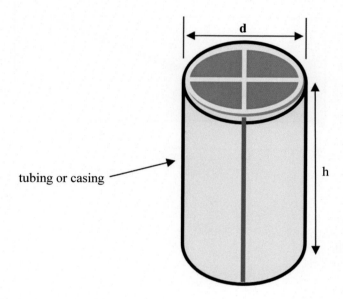

tubing or casing

200 ÷ by 0.0353 = 5,666 feet
h = the height of column of the fluid
A cylinder's volume = area × height
Area of cylinder = 0.7854 × d × d (the diameter)

All-Welded Wire-Wrapped Screens

Pipe Outside Diameter		Coupling OD	Screen Outside Diameter
2-$\frac{1}{16}$"	(2.0625")	Integral	2.61"
2-$\frac{3}{8}$"	(2.375")	2.875"	2.92"
2-$\frac{7}{8}$"	(2.875")	3.50"	3.42"
3-$\frac{1}{2}$"	(3.50")	4.250"	4.05"
4.0"	(4.00")	4.750"	4.55"
4-$\frac{1}{2}$"	(4.50")	5.000"	5.05"
5.0"	(5.00")	5.563"	5.55"
5-$\frac{1}{2}$"	(5.50")	6.05"	6.05"
7"	(7.00")	7.656"	7.55"

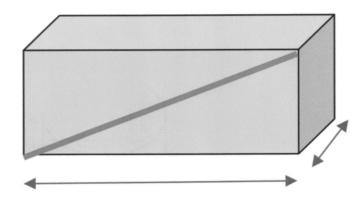

Rig Tanks:

volume = height × length × width
volume = 10' × 9' × 18' = 1,620 cubic foot
1,620 × 0.0178 = 288.5 barrels
1620 divided by 5.6 = 288.5 barrels

Rectangular rig tank capacity:

H = 10'
W = 9'
L = 18'

Preparing to Gravel Pack

Move in a workover rig to conduct prepack gravel packing.

Always think safety!

1. Move in the workover rig to the well's location as mentioned in previous pages.

2. Do not use a drilling rig to gravel pack! (Drilling rig personnel do not have sufficient knowledge about a cased-hole well operation and/or cased-hole gravel packing.)

3. A drilling rig is too expensive to operate, and the crew is not trained to do the gravel-packing task.

4. Move in and rig up a service rig on a solid, dry spot at the well, if you are using a land rig.

5. If working over the shallow bay water, load up equipment and pump up the rig barge and move the rig barge to the well's location using tugboats. (All the equipment must be secured to the barge before leaving the dock area.)

6. Spot the rig at the well. Fill up and sink the barge with bay water and level off the rig barge at the center of keyway. (Avoid running into the well while shifting the barge side to side.)

7. Drop anchors and rig up on the well. Fill up and level off the rig so that the wellbore locates in the middle of the keyway and the blocks hanging at the center of the well. (Check the flow lines before dropping or driving heavy anchors.)

8. Most flat barges do not have keyways. If using a flat barge, the rig must be tied to heavy-duty anchors welded onto the barge.

9. Use chains to secure all the tools and equipment that are subject to slide and fall into the bay.

10. Load up service rig and equipment onto the barge. (Chain down and secure the rig and equipment onto the welded anchors while tugging the barge in or out.)

11. Pump up the barge, move the barge with the service rig to the well platform, fill up and sink the barge, and level off. (Check to make sure there are no leaks in the barge and there is sufficient water in the bay to tug the barge safely.) Do not tug the barge in high wind, storm, or fog.

12. Dredge the channel, if necessary, before moving the rig barge to the well's site. Use shallow draft boats and barges (some barges and tugboats will draw too much water). Position the barge properly, drop anchors or tie the barge onto the pilings at the well's location, and level off the barge properly. Ensure the barge is leveled flat. Avoid running onto the well equipment and flow lines (lines are buried below the mud line). (Remember, zero-oil spillover in the water. One gallon of oil spill is reportable with fines.)

13. Rig up and hang the blocks at the center of the wellbore before tripping the production pipe.
Barge compartments must stay full to keep the barge stable. (The barge is subject to move with low tide and high tide.) Use multiple barges to transport other services, if necessary, to avoid hazards. Monitor weather and tide conditions at all times. Do not move a barge rig in a foggy day or night without proper guides. Check the overhead electric power lines in the bay before passing under any power lines. Review the tubing string condition, size, grade, and tubing rating.

14. Verify derrick rating and drill-line condition before working on the well to gravel pack.

15. Check the well location for hazards and overhead power lines (if any).

16. Move in and spot the fluid pump and a clean rig tank. Steam clean the rig tank and fill with clean water. Gravel-packing fluid must be filtered to 4 microns absolute.

17. Check and remove any fencing around the well prior to backing up to the well.

18. Use lock out and tag out if necessary. Install wind flags!

19. Do not rig up or rig down during the night (unless plenty of light is provided).

20. Do not rig up or rig down during high wind and storm.

21. If you are working on land, check the location to ensure the ground is dry and hard (suitable to rig up).
Use a steel mat to avoid sinking into the soft sand and clay near the wellhead.

22. All the rig crew must wear personal protective equipment, or PPEs (hard-toed boots, hard hats, eye safety glasses, working gloves, clean and dry long-sleeved shirt, and pants [no wedding rings, no neckless, and no earrings]).

23. Check for H_2S toxic gas. Never assume a well or location to be H_2S or CO_2 gas-free!

24. Start moving in and backing up to the well. If working on a land well, use two spotters to guide the rig backing to the wellhead. Do not run into the wellhead equipment and flow lines.

25. Rig up on the well. Hang the block/pipe elevators straight at the center of the wellbore.

26. Do not rig up the rig within twenty feet of any power line.

27. Test guy anchors 23,000 lbs. or as required. If using steel-based beams, the beam anchors and the guy-lines must be certified to stay in compliance.

28. No smoking within 150 feet from the wellhead.

29. String out guy-lines to the safety anchors, away from power lines.

30. Review the wellbore with a company man before you start working on the well.

31. Check and calibrate the weight indicator to avoid accidents.

32. Spot rig pump and a clean rig tank (wash and steam clean the rig tanks).

33. Fill up the rig tank with clean field-produced water or 2 percent KCL water.

34. Open the well. Read and record shut-in tubing and casing pressure.

35. Bleed off the well pressure down the flow line or into the rig tank.

36. Circulate to kill the well with brine if necessary. Bull-heading fluid down the casing and tubing string may not be the proper way to kill a well. The well may kick on you at any time later!

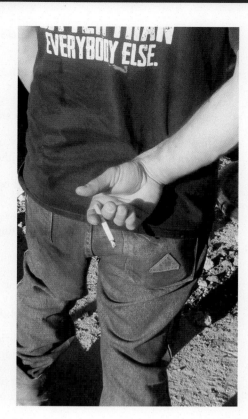

37. Unbeam rods, if any. Lay down horse heads away from the wellhead.

38. POOH and stand back sucker rods, if any or as recommended.

39. Nipple down the wellhead equipment and install blow-out preventers.

40. Test BOP 200 psi low and 3,000 psi high (never shortcut on the BOP testing).

41. Release or unseat the down-hole tools, such as tubing anchor catcher, packer, and seal assembly (if any).

42. Circulate the wellbore clean with brine or produced water to kill well and keep hole full of fluid. (The tubing string must be free of shallow holes and leaks before circulating it.)

43. Wellbore fluid must stay static during the workover and gravel-packing job.

44. Pull out of the hole with production equipment. Stand back tubing string, if any.

45. Keep the wellbore full with fluid properly to avoid well kicks (avoid pollution).

46. Never leave the wellbore open after pulling the production equipment out of a well!

47. Early detection and quick reaction will prevent accidents and major pollutions.

Run and Set an Isolation Bridge Plug as Foot Base

- A hard foot base is necessary to set the screen and liner in the casing string, below the open perforations.

- Rig up an electric wireline unit to run and set a cast-iron bridge plug.

- Nipple up and install casing lubricator or pack-off (as required). Never run wireline tools in a wellbore casing or tubing without a lubricator. (Do not take any chances.)

- Trip in the hole and drift casing and open perforation with a gauge ring and junk basket.
(Running a gauging ring and junk basket will provide information about the casing and perforation conditions before running the screen and liner.)

- Clear and drift the casing with a gauge ring to twenty feet below the open perforations.

- POOH and lay down the gauge ring and junk basket.

- Trip in the hole with a cast-iron bridge plug (CIBP) on an electric wireline.

- Correlate depth and set the cast-iron bridge plug twelve feet below the bottom set of open perforations. (Keep record of bridge-plug's landing depth.)

- The cast-iron bridge plug may be run and/or set on the tubing string at any desired depth if necessary. (You must keep accurate pipe measurements.)

- POOH and rig down the wireline tools and equipment.

Wellbore Preparation and Cleaning before Gravel Prepacking

a. Cased-hole wells always contain unwanted elements that need to be removed from the wellbore.

b. The old casing and tubing string are major nests for all sorts of deposit elements that must be removed before gravel packing (quality gravel packing).

Screen covered with mud

Screen covered with mud

c. The gravel-packing fluid slurry that is pumping downward in the casing-tubing annulus for several thousands of feet will scrape and remove rust, dust, paraffin, scale, and dehydrated drilling mud and will be mixed with your gravel-packing sand.

Debris will plug gravel-packing permeability tremendously and may land inside the screen.

d. Trip in the hole with rock bit and tandem casing scrapers to drift and scrape deposits in the casing string. Scale deposit, paraffin, dry mud cake, and all the contaminants must be scraped from the casing wall before gravel packing.

e. You may space out the casing scraper and the rock bit with a few joints of tubing while going in the well to avoid pushing the scraper through the open perforations. Use wire brushes to wipe off the casing wall.

Note: If the well is on partial vacuum, running a casing scraper will do more harm than good!

The scrapped solids will get into the formation and will be difficult to remove. Perforations must be isolated using a retrievable bridge (RBP), or cover up the perforations with gravel-packing sand before running a bit and scraper to remove solids from the casing.

f. There are several types of casing cleaning equipment available in the oil industry to brush off and scrape contaminated material out of the casing.

g. Never rotate or drill with scraper in a cased-hole wellbore. Scrapers are not designed to drill with!

h. Drilling with casing scraper inside a cased-hole may cut and damage the casing or cause damages to the scraper.

i. Continue going into the well with the bit and scraper to the top of the open perforations. Do not force the scraper through the tight casing spots; the scraper may go through tight spots, but it will be difficult to come back out!

j. Circulate the wellbore clean with filtered field saltwater (circulate well on the short way).

k. Pull out of the hole with tubing string; lay down the bit and scraper assembly.

The Concept of Using a Perforation Washer in Wellbore Cleaning

Using a Perforation Washer

- Make sure to keep good tubing measurements.

- Make sure that the tubing string is free from leaks or corrosion holes.

- Space out the isolation cups on the tool to wash perforations one foot or two feet of perforation at a time, or as needed.

- Trip in the hole with perforation washer tools slowly to avoid surging fluid.

- Clear open perforations and lower perforation washer just two feet below the open perforations.

- Break circulation, pumping down the tubing string and out through the tool. Pick up and wash the perforations one foot at a time, pumping fluid down the tubing string and out of the tool's ported hole.

Tubing tester unit

- When pumping down the tubing string, the brass ball at the bottom of the tool will be seated, and fluid will be forced out of the ports between the cups and out into the annulus through the casing's perforated holes.

- Continue washing the perforations one foot at a time, and break down the perforations at two to three barrels per minute for a duration of seven minutes for each set of perforations.

- Continue washing and circulating down the tubing and out of the annulus until all the perforations interval are broken down and cleaned up to satisfaction.

- Shut down and reverse circulate the well clean, if necessary.

- The tool will be closed or opened by right-hand rotation to wash and break down each set of perforation.

- Spacing out the tool at one foot at a time will allow you to use either acid or plain water to break down perforations.

- After washing the open perforations, the tool can be opened and lowered to the rathole below the open perforations to reverse circulate the debris of sand and solids out of the wellbore.

- Application of perforation washer is unique by pumping water or acid down the tubing, through the tool, and into the perforated tunnels at the rate of two, three, or four barrels per minute. Pressure rate of 1,000 to 2,000 psi can be applied to break down and remove mud cakes and break-down the tight perforations.

- Slow down while pulling the perforation washing tool out to avoid swabbing the formation sand back into the wellbore.

- Fluid loss into the formation will be noted while washing and circulating fluid. Loss of fluid may be an indication that the formation will take a good volume of gravel-packing sand.

- Very little loss of fluid into the perforation may be an indication of tight formation or plugged-off open perforated tunnels. (Clay and shale bed can swell and block off the sand pack.)

- Pull out of the well with the perforation washer and prepare for sand prepacking.

Note to Operators

If perforation washing tools are not available, consider acidized swab and circulate the wellbore clean to remove contaminated drilling mud, iron sulfide, shale, and clay out of the invasion zone.

During the drilling operation, a considerable amount of drilling mud may be lost into the productive reservoir because of higher differential pressure, greater than the formation pressure.

The mud filter cake must be removed from the perforation tunnels before any type of gravel packing. Drilling mud, mud filter cake, and lost circulation material (LCM) are bad objects when they mix with resieved gravel-packing sand. Contaminated material will reduce productivity!

If perforation washer is not available, you may consider a casing swabbing to pull contaminated drill mud and LCM pills out of the formation face. (Dummy guys have lost all kinds of bad stuff into the hole.)

Wellbore Inspection and Preparation before Gravel Packing

1. Make a final inspection of tools and equipment (quality first).

2. Check the rig pump and tanks. Make sure the fluid tank is very clean and with sufficient volume of water available during gravel-packing operations. (Do not run out of water during gravel packing!)

3. Note: Some newly completed wells are alive and may flow during gravel packing.
   Heavy filtered brine water may be used to balance the reservoir pressure and continue to complete the gravel packing successfully (Quarantine Bay, Ship Shoal, Block 198, and south Louisiana).

4. Move in and spot the filter unit. Use fluid filters as required (2 to 6 microns fluid filters are acceptable). Change fluid filters more often to reduce back pressure and avoid pumping dirty fluid with gravel-packing sand!

5. Trip in the hole with the rock bit and tandem scrapers. Scrape and drift the casing string to the top of the open perforations. Use tandem wire brushes to rotate and clean up the casing string as best as possible. Circulate wellbore clean using filtered brine fluid.

6. The application of acid pickling in contaminated casing and tubing is recommended to remove viscous drilling mud and gel pills out of the wellbore (see acid pickling procedures).

7. POOH and lay down bit and scrapers.

Casing Inspection Test before Gravel Packing (Mechanical Integrity Test)

1. Trip in the hole with isolation packer to the top of the casing perforated holes.

2. Set the packer and test casing string above the open perforations.

3. Test the casing string as high as recommended before gravel packing.

4. Take extra precautions when testing the old casing wellbore (some casings may break down during mechanical integrity test [MIT]).

5. If the casing string failed to test, stop the gravel-packing operation (prepare to repair casing).

6. If the casing string tested good, proceed ahead!

7. POOH with work string; lay down squeeze packer. Keep the hole full constantly!

8. Check the well for flow. If the well is not flowing, continue to the next step.

9. Rig up the wireline tools and equipment. Nipple up and test the casing lubricator. (Never use wireline equipment without a lubricator.)

10. Trip in the hole with a gauge ring and junk basket. Clear and drift the casing to twenty feet below the open perforations.

11. POOH and lay down the gauge ring and junk basket. (Running a gauge ring and bridge plug through is a positive step to run the screen and liner in the well without difficulty. Running the gauge ring will ensure that the bridge plug and screen/liner will pass through the casing and the open perforations without problems. (Do not shortcut. I already did it for you!)

12. Trip in the hole with a cast-iron bridge plug. Correlate and set the bridge plug eight to twelve feet below the open perforations. (The purpose of the cast-iron bridge plug is to establish foot base for the screen and liner assembly to set on!)

13. Note: if there is a concern that the cast-iron bridge plug may leak, you may consider placing cement on the cast-iron bridge plug using a cement bailer.

14. I do not recommend setting a screen and liner on top of the formation-sand bridge. If the sand bridge can be used as a bridge plug, top off the sand with cement to obtain a hard bottom.

15. POOH and rig down the wireline tools and equipment.

16. A gravel-packing company will not be responsible for the wellbore preparations.

17. The screen manufacturing company will provide a schematic of the screen/liner and will forward it to the well operator for approval (based on the information provided to them).

Screen and liner schematic will basically show the following:

1. The screen and liner size is referred to as the outside diameter of the screen (the circumference of the screen wrap and the outside diameter of liner tub). Check the screen and liner schematic very carefully before it is made up!

2. For all practical purposes, the space between the inside diameter of the casing (casing ID) and the outside diameter of the screen (screen OD) must be at least two inches or larger.

3. The larger the space between the casing and the screen annulus are, the more resieved gravel can be placed and packed between screen and the gun-perforated holes.

4. The length of the screen section—a good screen design will cover at least five feet of screen below the open perforation and five feet of screen above the top perforations.

5. Do not shortcut on the screen length. (I've already done that for you.)

6. The screen slot size is the gap between the screen wraps. The design of a screen gap is based on accurate formation-sand analysis. *Screen gauge* is referred to as "the opening between the wire wraps." (An accurate slot opening is important in gravel packing success.) Force the selected sand gravel with your fingertips into the screen slots. The selected gravel-packing sand must not go through the screen slot (0.012 gauged and 20/40 sand).

7. The selection of screen and liner is often based on formation-sand samples, the rule of thumb, and practical local experience!

8. The selected resieved gravel-packing sand should be six to eight times larger than the small analyzed formation-sand grain. The screen gap is designed based on the 10 to 20 percent of the resieved sand samples taken from the formation.

9. Liner section. The blank liner section is made up of several joints of carbon steel tubing of various sizes, referred to as blank pipes, that extends above the screen section. Blank joints are used to convey the

screen/screens in the well and also hold gravel-packing sand in the annulus above the screen section. The height of gravel-packing sand around the liner will maintain constant sand cover over the perforations.

10. The liner section may extend sixty to four hundred feet based on the gravel-packing application.

****

B. The concept and ideas behind the prepack gravel packing

— The application of resieved prepack gravel-packing sand into the open perforation tunnels is the favorable and ideal technique to place high-porosity gravel across the reservoir formation and keep the unconsolidated sand under tight gravel-packing control.

— Pumping gravel-packing sand down the tubing and out into the perforation and formation tunnels will block off the migration of fine formation sand and solids into the wellbore. You may pump and sand-pack max quality and quantity of gravel-packing sand behind the casing.

Prepack gravel packing is conducted in seven steps:

1. Trip in the hole with thirty feet mule shoe joint; wash and clean up the wellbore.

2. Pump sand and prepack gravel into the formation through the perforations as much as possible.

3. Run and set an elected screen and liner at the bottom.

4. Pump and displace the required volume of clean gravel-packing sand around the screen and liner.

5. Release from the screen and liner.

6. Run and set an isolated packer over the screen and liner.

7. Put the well on production.

****

Presenting a Prepack Gravel-Packing Method

Prepacking is the first step to Braden Head squeezing or forcing gravel-packing sand out through the open perforation tunnels and into the formation using plain, clean, filtered saltwater.

The concept of prepacking is to use low pressure sand-squeeze pack below the formation fracture pressure to force gravel sand through the casing perforation tunnels and behind the casing at a laminar flow to obtain a tight gravel-packing distribution.

The resieved gravel will have higher porosity and higher permeability than most of reservoir formation-sand grains. Placing resieved sand gravel at the face of the formation will reduce fluid flow velocity and prevent formation from disturbances.

If the formation is taking water, it will take gravel-packing sand for certain.
If the formation is not taking water and/or is taking in a small amount of water, it will not take too much prepacked gravel sand.

- Stop disturbing the reservoir formation by using high-pressure hydraulic pumps.
- Avoid mixing and contaminating the gravel-packing sand with formation solids.
- Do not rearrange the formation sand behind the casing string; it will reduce productivity.

Prepack gravel-packing method is done in several steps:

Step 1. Wash and clean up the wellbore before prepacking the open perforations.

Step 2. Prepacking. The purpose of prepacking sand gravel is to pump, displace, and pack clean resieved sand gravel into the perforation tunnels and behind the casing onto the formation using an open-ended mule shoe joint. Placing gravel-packing sand is the number 1 step of holding formation sand behind casing.

Step 3. After a successful prepacking, the wellbore fluid and contaminated sand will be circulated out.

Step 4. Run and set a selected screen and liner in the place, and pump gravel-packing sand around the screen and liner assembly (from the bottom of screen up to the telltale screen section).

Step 5. Run and set an isolation packer.

Step 6. Run the production equipment, and place the well on production.

Note: after the prepacking is done, the wellbore condition will dictate to you on how to run the screen and liner in the well and gravel pack the screen/liner in place.

The choice and selection of procedures are based on the wellbore conditions.

****

Prepacking Sand Gravel Using a Mule Shoe Joint

Prepare the rig equipment and handling tools to conduct prepacking.

1. Move in and spot a clean fluid tank or tanks on the location.

2. Steam clean and fill up the rig tanks with clean and clear water. (Keep plenty of water on the location.) Make certain that the fluid tank is clean and free from dirt, paraffin, oil sludge, drilling mud, and harmful elements carrying bacteria. You must remove mud, sand, silt, and iron rust from the tanks, as well as dirty water with iron sulfide. Iron oxide and bacteria must not be used in gravel packing.

3. Spot and rig up the gravel-packing unit on a leveled ground near the wellhead.

4. Spot the filter unit on a flat and leveled ground near the gravel-packing unit and fluid tank.

5. Check and install new filters. Discard all the used filter elements.

6. Install large-sized rubber suction lines from the storage water tanks to the filter unit/units.

7. Install discharge lines from the filter unit to the gravel-packing unit (prefer steel lines).

8. Rig up the fluid manifold. Install steel flow lines from the manifold to the casing and from the manifold to the tubing string and from the circulation return line onto the rig tank.

9. Wash and pickle the lines and equipment to remove old contaminated elements.

10. Check to make sure the pumping lines are installed correctly, going in the right direction. Double-check the lines and the valves before gravel packing. It is easy to make mistakes pumping the gravel the wrong way. (It has happened before!)

11. Install the flow meter. (Flow meters are a great help in fluid measures and displacements.) Establish the pumping rate through the filter system to make sure the fluid meter is functioning and working properly and all the lines are clean and fully open.

Cut and prepare a mule shoe joint (single or dual mule shoe).

Going in the Hole to Conduct Prepack Sand-Gravel

Note: I would prefer to prepack and run the screen and liner in the same day if possible.

- If there is not enough time to prepack the sand gravel and run the screen and liner in the well on the same day, then the prepacked perforations must be covered with fresh, clean sand gravel to protect the prepacked formation and prevent contaminants from entering into the wellbore.

- Spot and rig up the gravel-packing tools and equipment. Spot the required volume of gravel-packing sand sacks near the gravel-packing unit (to get started).

- Check the gravel-packing sand to make sure it is the correct size and grade. Do not mix gravel-packing sand (20/40 mesh sand with the 15/20 mesh).

- Do not pump mixed gravel-packing sand into the well (uniform gravel size is required).

- Trip in the hole with a thirty-one feet mule shoe joint on the work string open-ended tubing. (Avoid using a sharp-pointed mule shoe joint. Sharp mule shoe may hang up.)

- Test the tubing string at 4,000 psi going into the well. (The work string must be in good working condition, clean, and without holes!)

- The work string must be clean and free from paraffin and formation scales (rabbit and drift the tubing string to ensure a full drift.)

- Measure and keep an accurate pipe tally. Keep an accurate count and the tally of all the pipes on the location to avoid miscount. (Do not keep too many joints scattered around the well's location.)
  Finish testing and going in the hole with the work string and mule shoe joint.

- Stop one hundred feet above the open perforations to check the string weight. Break circulation and check the casing integrity by circulating fluid.

- Rig up a circulating type or power swivel. (I do not recommend using a wing-union swivel). Fill up the hole with fluid to break circulation. Check and reverse circulate wellbore fluid (circulate bottoms up while catching fluid samples). If you find sand, mud, or solids in circulation, casing integrity may be questionable. If the fluid circulations turn out to be clean, the casing condition may be good, so continue on to the next step.

- Continue washing and going down and tag bottom on the bridge plug to make sure the bridge plug is still there. Use the bridge-plug depth as the measuring depth reference.

- Wash and reverse circulate the wellbore to the top of the cast-iron plug, below the open perforations. Correlate the tubing measurement with the bridge-plug depth as reference.

- Do not drill or spud on the bridge plug with a mule shoe. Break circulation and check for fluid loss and formation solids in the returns. (If the well is taking some fluid, it is a good indication for a successful prepack gravel packing.)

- Keep the fluid hydrostatic slightly higher than the reservoir pressure.

- In the Gulf Coast area, the shallow reservoir formation produces sticky solids from shallow casing leaks. The sticky formation material will pack off with high pressure and release solids at a low circulating pressure. (Watch carefully for sticky clay and shale.)

- All the wells normally communicate information to you, so make sure you understand that and take advantage of the well information to stay safe and to cut cost! When people say that the well came in all of the sudden, it is not a true statement. (Wellbore fluid will give you plenty of warning before kicking!)

- Before prepacking commencement, pull the mule shoe joint two hundred feet above the perforations. Reverse circulate fluid at a low pressure and low rate at least two tubing volumes while circulating with filtered water and catching fluid samples. (Use socks to catch samples.)

- If you recover any kind of formation solids, such as red/blue clay, dehydrated mud, shale strikes, and sand or cement, it is a good indication of a leak or leaks in the casing string.

- Never surge pressure during gravel-packing work. You may break down the casing corrosion pits! (Always bleed off pressure slowly or allow the pressure decline by itself.)

- If there is no indication of solids or leaks in the casing string, continue forward.

- Rig up the gravel-packing unit and prepare to mix and pump gravel. (Make sure sufficient volume of water and gravel sand is available to complete the job.)

- Conduct a safety meeting and explain work procedure before starting gravel packing!

- Check the lines and valves once again for correct pumping direction of fluid flow to the well. Avoid pumping the sand slurry in the wrong direction!

  I prefer rigging up a stripper-head on the BOP stack to reciprocate the tubing string at this point. Always work with a circulating swivel and/or power swivel.

- Start mixing sand. Pump and displace sand gravel at low concentration down the tubing while watching fluid circulation coming out of the annulus into the rig tank. Depending upon the tubing and casing size, you may mix and displace two sacks to four sacks of gravel at a time, or as needed, while watching the displacing volume and the time closely.

- Once the sand gravel reaches near the bottom, start reciprocating and moving the mule shoe up and down twenty-five feet or longer at a constant slow speed from the bottom of perforations to the top of perforation, up and down.

  As soon as the sand reaches to the bottom, close the casing valve and the blowout preventers at the surface while Braden-squeezing the sand gravel out through the open perforation tunnels by working the pipe and displacing the gravel down the hole. (Use a stripper head to assist reciprocating the pipe and to avoid damaging the rams in the BOP stack.)

- Continue working the pipe while pumping the gravel-packing sand away through the perforations. Do not stop moving the pipe to prevent early sand-out. (If the well is taking sand and fluid, continue.)

- Pump and displace sand into the formation at a low concentration as much as practically possible. There is a large void behind the casing in the old wells that produced formation sand for many years. The void space must be filled with new gravel-packing sand.

- Squeeze the sand gravel through the perforation tunnels and behind the casing at low sand concentration while recording the initial packing pressure through the open perforation.

- Continue pumping and displacing sand gravel into the formation at low-squeeze pressure not exceeding 1,100 psi prepacking pressure, or as directed.

- Prepacking sand into the formation is an impressive and encouraging method of placing gravel.

- Pumping gravel in nonviscous fluid can be accomplish at $\frac{1}{2}$ to 3 pounds/gallon of fluid at a rate of two to three barrels of completion fluid.

- Record the initial gravel-packing pressure and the final prepacking pressure. (If for some reason the perforations become covered with sand pack, you may have to wash and reverse out the sand and start prepacking again to obtain the maximum anticipated prepacking sand volume and pressure.)

- In case of prepacking several perforated intervals in the same wellbore, the prepacking will start from the top perforated interval first.

- Do the top perforated interval first, then second, and finally, the third set of perforations. Always prepack from the top down to prevent getting the pipe stuck.

- During the course of prepacking, the casing valves and blowout preventer will be closed shut to force the gravel out through the perforation (Braden squeeze). The tubing and the casing string will be subject to the prepacking sand-squeeze pressure.

- The gravel distribution will occur through the less-resistant part of the perforation area first and then distribute to the tighter area. Continue pumping sand and reciprocate tubing until the anticipated sand-out or predetermined higher-pressure squeeze pack is reached.

- Once a sufficient volume of gravel is pumped into the perforated tunnels and out into the formation behind the casing, the prepacking pump pressure will start increasing to a pressure pack higher than initial packing pressure.

- Prepack the gravel to the anticipated packing pressure as required. Record the volume of sand and water out into the formation.
  Prepacking one thousand to two thousand pounds of gravel is considered as low volume pack (tight formation).
  Prepacking two thousand to four thousand pounds of gravel is considered as average water pack.
  Prepacking four thousand to six thousand pounds of gravel and over is considered as high prepack.

- If the gravel prepacking is satisfactory, pull the mule shoe in clear and wait for thirty minutes for the prepacked sand to set in place behind the perforations. Slack off, wash down, and reverse circulate the excess gravel-packing sand out of the wellbore at a slow circulating rate to the top of the bridge plug. (Avoid lifting heavy loads; it will cause loss of circulation.) Circulate the well clean at low pressure and low rate. (Use heavy filtered brine, if necessary.)

- Pull the mule shoe in clear above the perforations. Shut down the operation and wait for at least one hour while keeping the hole full with water (do not pump the prepacked sand away).

- Lower the tubing and retag the bottom to find out if any sand or solids are entering the wellbore. Wash and reverse circulate the excess sand and solids out of the rathole, if any. Catch samples to check for sand intermix.

- Pull the mule shoe in clear and shut down. Wait for one hour while keeping the hole full of fluid, if necessary, while preparing the screen and liner and all the necessary tools and equipment.

- After waiting one hour, lower the tubing string and mule shoe joint. Tag the bottom at the bridge plug and mark the pipe at the rig floor for measurement. If the well is taking fluid, the wellbore may be stable for tripping out and tripping back into the well with the screen and liner using a crossover tool.

- If the wellbore stays clean and stable, start out the hole while keeping the casing completely full of fluid. Continue keeping the well full of fluid to maintain hydrostatic head on the prepacked gravel until it is ready to run the screen and liner in the well. Pull out of the well slowly to avoid swabbing the sand into the wellbore.

After the successful gravel prepacking, the wellbore will dictate two choices to you on how to run and set the screen and liner in the well and gravel packing it in place:

Choice I

If the wellbore stays clean and stable after prepacking, you have a choice of running the screen and liner in the hole and gravel packing in place using a crossover tool (x-over tools).

Using a crossover tool is a quick and an efficient method of landing the screen and liner. It is the best choice of landing the screen and liner in the well. (Stick with it.)

Choice II

If the wellbore is unstable after prepacking and flowing back sand and solids, you must cover the open perforations with gravel-packing sand, wash down, and place the screen and liner to the bottom and pump gravel-packing sand around the screen and liner.

Try to run the screen and liner without the washing-down method, if you can (it is a lengthy process).

Choice I—Stable Wellbore for Gravel Packing

Running the Screen and Liner Using a Crossover Tool

The crossover packer is an excellent tool for a single-stage gravel-packing purpose.

The application of a crossover tool will keep gravel slurry contained inside the tubing string and minimize contamination of the gravel-packing fluid. The application of x-over packer is an efficient and a problem-free technique, if it is done correctly.

A crossover tool consists of inner and outer sleeve mandrels with ported holes.

The tool is built with two isolation casing rubber/wire swab cups facing down with a tight fit against the casing string to direct fluid and gravel-packing sand downward.

Sand and carrying fluid will be pumped down the work string and through the crossover tool and into the annulus space between the casing and outside of the screen and liner assembly.

Fluid and sand slurry will pass down the annulus. The sand will fall and pack around the screen and liner while the carrying fluid returns through the screen and up through the x-over tool above the isolation cups into the annulus and up the hole (very simple, unique, and quick).

No mechanical rotation is needed to set and/or to release the x-over tool.

Make sure to have an accurate measurement of anything that goes into the hole.

Lay down extra joints of tubing to replace the length of screen and liner joints.

Trip in the hole with the screen and liner assembly with the crossover tool (x-over tool).

- Bull plug (drag bull plug and/or flat bottom bull plug).
- Screen sections (epoxy-coated inside and outside of base pipe).
- Blank liner joints (epoxy-coated internally and externally).
- Telltale joint (epoxy-coated internally and externally).
- Left-hand polished nipple (coated internally and externally).
- Right-hand release back-off sub (tool will be retrieved out of the hole after gravel packing).
- Crossover tool will be retrieved out of the hole after gravel packing.
- Test and internally clean the work string tubing.

    Snug tight the back-off sub into the polished nipple (to avoid too much torque).

    ► Trip in the well with a screen and liner assembly and work the string slowly (while pumping water going down the annulus at a slow rate).

    ► Slow down going in the well to avoid fluid surging because of the crossover rubber cups going in the hole. (In the first few stands, you may have to force the crossover tool down the casing because of friction and lack of weight.)

    ► Run and tag bottom on the bridge plug (footing).

    ► Pick up six inches off the bridge plug and rig up the gravel-packing unit and equipment quickly. (All the gravel-packing equipment should be ready at this point!)

    ► Break circulation down the tubing and out of the crossover tool and out of the screen and liner to the surface into the rig tank.

    (Fluid will travel through the tubing string and the crossover tool into the annulus between the casing and the screen and liner. Water will return through the screen into the x-over tool and up the annulus to the surface.

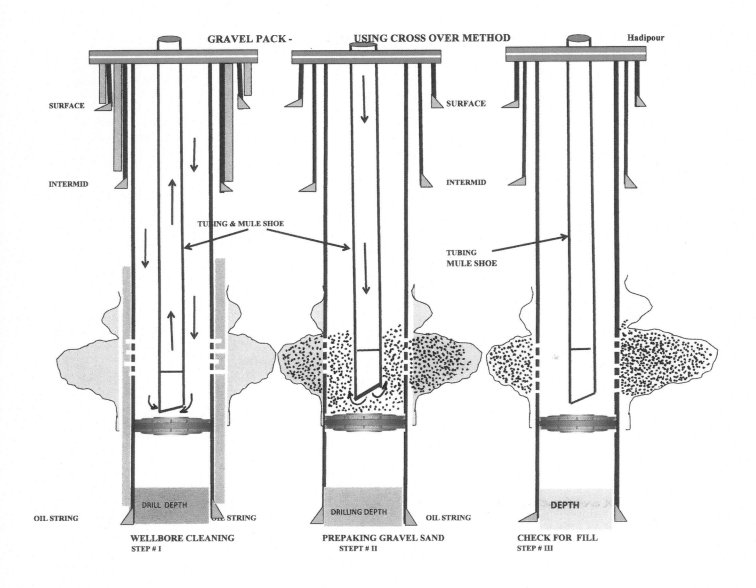

GRAVEL PACK -   USING CROSS OVER METHOD   Hadipour

WELLBORE CLEANING
STEP # I

PREPAKING GRAVEL SAND
STEPT # II

CHECK FOR FILL
STEP # III

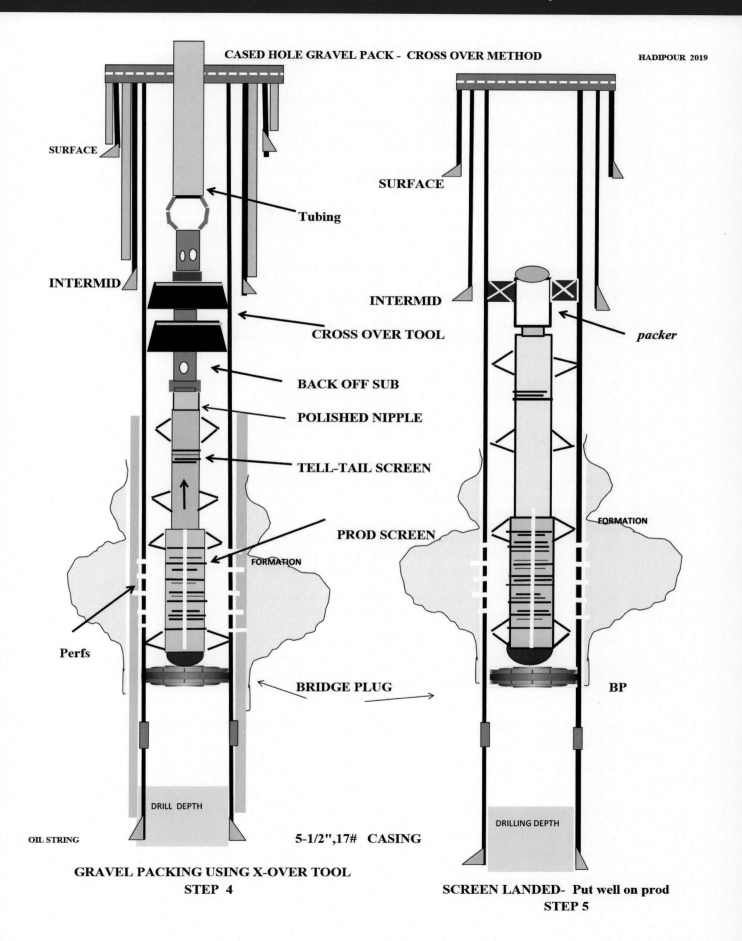

CASED HOLE GRAVEL PACK - CROSS OVER METHOD

HADIPOUR 2019

SURFACE

Tubing

INTERMID

CROSS OVER TOOL

BACK OFF SUB

POLISHED NIPPLE

TELL-TAIL SCREEN

PROD SCREEN

FORMATION

Perfs

BRIDGE PLUG

DRILL DEPTH

OIL STRING

5-1/2",17# CASING

GRAVEL PACKING USING X-OVER TOOL
STEP 4

SURFACE

INTERMID

packer

FORMATION

BP

DRILLING DEPTH

SCREEN LANDED- Put well on prod
STEP 5

105

- Mix, pump, and displace the calculated volume of the required sand down the tubing string through the crossover tool and around the screen and liner in the annulus and up to telltale screen.

- The casing annulus will fully be opened to the rig tank for the circulation returns.

- Displace gravel at a slow rate and with low sand concentration to prevent sand bridging while pumping through the tubing string and the small outlet in the crossover sub.

- Continue pumping and displacing sand until resieved gravel sand reaches up to the short telltale screen.

- Once the gravel-packing sand reaches the telltale screen, you will see the increased pump pressure and decrease on fluid circulation coming out into the rig tank. (This is normal.)

- Continue pumping at a slow rate and stress the sand pack at 700 psi or as required.

- Shut down and wait for sixty minutes, allowing the sand to pack off, set, and heal.

- Re-stress the sand pack at 700 psi again (or as requested). If the sand pack's pressure declines and circulation increases, you may add more sand, if necessary. Adding more sand is a professional judgment call.

- Never pump too much sand at one time to bring the gravel-packing sand up to the telltale screen. You may need only a half sack of sand at a time.

- If the sand gravel packing is successful, pick up and pull upstream on the tubing string at eight hundred to one thousand pounds tension slowly. (Avoid pulling the screen up the hole.)

- Right-hand rotate the tubing string for eight to twelve rounds at the tool with a pipe wrench to release the back-off sub from the polished nipple. (Often you may see the pipe jump free because of tension.)

- Pick up six inches off the polished nipple and reverse circulate the wellbore to clean (two volumes). Catch samples at the rig tank. You should not see any gravel-packing sand in the return. (You've done it; you are now free and clear from worry!)

- POOH with the tubing string slowly, and lay down x-over tools and the back-off sub. (The wellbore should be protected with gravel-packing screen and liner.)

- Trip in the hole with an isolation packer (either an O-ring type or a lead-seal type). The choice of isolation tools will depend on the company man supervising the project.

- If the remedial gravel packing is done successfully, trip in the hole with production equipment and put the well on production.

- Nipple down and install the wellhead equipment and flow lines. Turn the well over to production.

- Always bring the well in slowly after gravel packing. Allow the sand to set and heal properly.

- Rig down the service rig, clean up the location, and move out.

- Haul off trash, food wrappers, dirty gloves, fluid filters, plastic sacks, etc. and take it with you.

- Keep the environment clean!

Choice II—Unstable Wellbore

Wash-Down Method of Running the Screen and Liner in Place

1. After sand prepacking with the mule shoe is completed, wash and reverse out the excess gravel-packing sand out of the wellbore to the bridge plug.

2. Pull the mule shoe joint at three hundred feet above the perforations and wait for at least one and a half hours. (You may keep the hole full the entire time slowly.)

3. Shut down and check for flow.

4. If the well stays full of fluid and/or runs fluid over, the wellbore may not be stable, and the screen and liner must be washed in place.

5. Slack off the tubing and mule shoe joint to check and tag the bottom.

6. If the prepacked sand with formation sand is caving into the wellbore above the bridge plug, the well may be unstable. You must wash the screen and liner in place.

Wellbore Must Be Prepared before Wash-Down Method

The wash-down method is done when prepacking sand flows or falls back into the wellbore, causing the hole to become unstable to run the screen and line.

This type of well will normally take a little water during prepacking and will flow fluid back because of reservoir pressure or unstable cavity behind the casing. Any amount of fluid flowing back is sufficient reason for washing the screen and liner in place to avoid repeating trips in and out of the well to get the screen and liner in place.

When formation-sand gravel caves into the wellbore, it will displace fluid out to the surface.

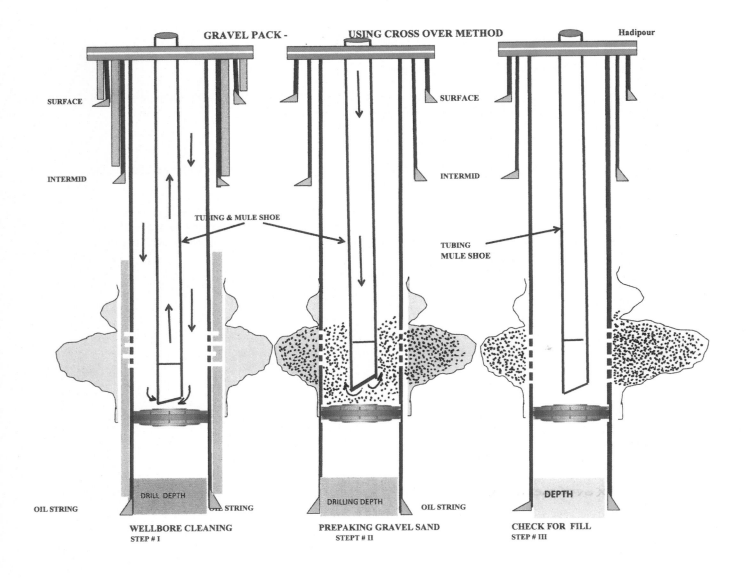

GRAVEL PACK - USING CROSS OVER METHOD

Hadipour

SURFACE

INTERMID

TUBING & MULE SHOE

SURFACE

INTERMID

TUBING
MULE SHOE

OIL STRING

DRILL DEPTH

OIL STRING

DRILLING DEPTH

OIL STRING

DEPTH

WELLBORE CLEANING
STEP # I

PREPAKING GRAVEL SAND
STEPT # II

CHECK FOR FILL
STEP # III

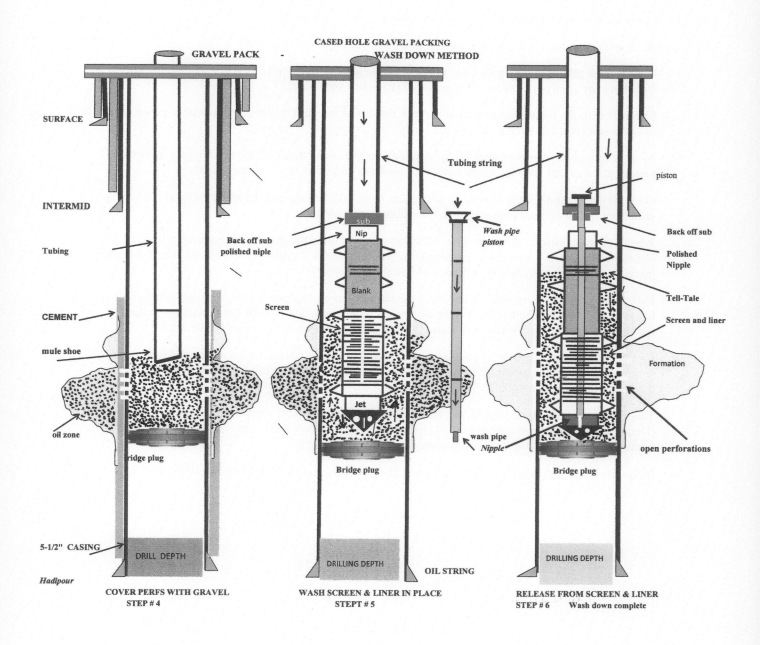

CASED HOLE GRAVEL PACKING
WASH DOWN METHOD

GRAVEL PACK

SURFACE

INTERMID

Tubing

CEMENT

mule shoe

oil zone

ridge plug

5-1/2" CASING

Hadipour

DRILL DEPTH

COVER PERFS WITH GRAVEL
STEP # 4

Back off sub
polished niple

Screen

Bridge plug

DRILLING DEPTH

WASH SCREEN & LINER IN PLACE
STEPT # 5

Tubing string

sub
Nip

Blank

Jet

Wash pipe
piston

wash pipe
Nipple

OIL STRING

piston

Back off sub

Polished
Nipple

Tell-Tale

Screen and liner

Formation

open perforations

Bridge plug

DRILLING DEPTH

RELEASE FROM SCREEN & LINER
STEP # 6   Wash down complete

The Wash-Down Method

(Avoid wash-down method as much as possible!)

- In order to stop the prepacked gravel sand from flowing back into the wellbore, one must trip back into the hole with thirty-one feet of mule shoe joint.

- Wash and circulate the wellbore clean using balanced heavy water to keep fluid flow under control.

- (Keep the hydrostatic head slightly more than the reservoir pressure.)

- Wash and reverse circulate the wellbore clean with filtered brine, if necessary.

- Displace and fill up the hole with clean gravel-packing sand from the top of the bridge plug up to thirty feet above the open perforations, or as required (balanced sand plug).

- Slack off with the mule shoe joint, and recheck the top of the sand bridge above the perforation for a record and sand-calculation requirement.

- Note: keep a good record of sand-bridge depth above the perforations before washing the screen and liner in place so that you may know how much more, or less, sand you may need later.

  Once the screen and liner is washed in place, you must calculate the needed additional gravel-packing sand from the above sand plug and up to the telltale screen section.

  When you wash down a screen and liner through the sand plug, the gravel-packing sand will be lifted higher because of the screen section occupying some of the sand space area. (Assume that it did not pump or did not lose any amount of the sand into the formation.)

- Pull up and wait for thirty minutes and retag the sand plug above the perforation to make sure no influx is coming into the wellbore. (The weight of the sand plug is sufficient to hold the well under control.) If the gravel plug is stable, POOH with the mule shoe joint slowly while keeping the hole full.

- Lay down the mule shoe joint and extra joints of tubing in singles to replace the screen and liner joints' length. Remeasure the screen and liner joints. Mark and space the last joint to properly tag the top of the bridge plug's landing point.

  The gravel operator should know the exact tubing measurements to reach and tag the bridge plug. (Double-check the pipe counts and tally before starting in the hole with screen.) You may have to space with pup joints to land the work string at the rig floor.

- Never work above the rig floor higher than five feet for safety and well-control purposes!

Start in the Hole with the Screen and Liner and Wash-Down Tools

- Make up the wash-down set shoe at the bottom of the screen section.
  Check the wash-down shoe to make sure it is functioning properly. The wash-down shoe is made with a spring-loaded ball and seat that works with the pump pressure to allow the cleaning fluid to pass through. When the pressure is bled off, the tool will be closed shut automatically, preventing sand and solids from getting into the screen from the bottom.

- Start in the hole with the centralized screen assembly section, and the wash-down shoe, followed by the centralized blank liner joints (as many as required and spaced fifteen feet apart), the telltale screen with a blank pipe (the down-hole camera), and the left-hand polished sub (either machined polished or internally and externally coated nipple).

- The right-hand releasing tool (short x-over) must be snug and not too tight.

- Space the small-sized wash pipe joints (slick OD with strong and stable acme threads).

- Rubber piston cup–type tool as a packing-off element.

- Correct-sized seating nipple above the rubber piston packing-off tool.

- The work string size (2-⅜″, 2-⅞″, or 3-½″, etc.).

- Trip in the hole with the screen and liner and the wash-down tubing inside of the screen and liner.

- Tag the sand bridge above the perforations, mark the work string at the floor, check the tagging depth with the pipe, and space with necessary pup joints to avoid shutting down during the wash down. (It must be spaced to avoid mistakes of shutting down or getting stuck.)

- Pick up ten feet above the sand pack, and check the depth spacing again (critical steps).

- Make sure everything is in order before you start washing, including water and fuel—you don't want to run out of these.

- Break the fluid circulation down the tubing string through the small wash pipe, out of the wash shoe or jet shoe, and up the annulus (casing annulus will fully be opened).

- Check the fluid circulation, rate, and pressure. (You may not stop to correct anything after this exiting point!)

- Increase the pumping rate and pump pressure to a comfortable circulating condition suitable to lift the gravel-sand to get it down. (It would take a few minutes to get the job done.)

- Lower down the wash tool (set shoe). Wash through the gravel pack while going down at a constant speed and good circulating rate of 2.5 to 3 barrels per minute without interruption (washing and lifting gravel-packing sand up the annulus). This critical wash-down step must be carried out for only a few minutes in one continuous motion without shutting down.

- By washing down with the shoe into the sand, you actually stir and lift the entire gravel-packing sand plug up the annulus with the jetting action of the fluid coming out of the shoe.
  (The annulus is the space between the casing string and the screen and liner assembly.)

- Once the wash shoe reaches the tag on the bridge plug below the open perforations, your pipe mark should be at the expected point at the slips on the rig floor. (You are made it!)

- Shut down the circulating pump and allow the gravel-packing sand to fall back around the screen and part of the blank liner. (This entire wash down will take only a few crucial minutes if done correctly.)

- When you are circulating and lifting the gravel-packing sand, you might also pump some of your sand into the formation.

- Pump a few barrels of filtered water down the annulus now to chase the sand grains off the casing collars and to make sure the sand grains are pushed back in landing position. (You should know the new top of the gravel in the annulus across the perforations at this point.)

- The amount of extra gravel-packing sand is computed by volumetric calculation to raise the gravel-packing sand up to the telltale screen section.

- You must be careful not to dump too much sand at once to fill up the void from top of the sand to the telltale screen!

- Continue pumping and displacing sand gravel down the annulus while pumping and circulating filtered freshwater. Adding and pumping sand-gravel down the annulus is similar to poor-boy gravel packing.

- (You may reach a point that you have to pump half a sack of sand at a time to avoid mistakes of pumping too much sand.) Take your time to avoid mistakes.

    Mistakes are aggravating punishments, mentally, bodily, and financially. (Avoid it if you can.)

- Once the gravel-packing sand reaches the telltale screen section, the pump pressure in the annulus will increase and the circulation rate will decrease, indicating a sand pack around the screen and liner from the top of the bridge plug up to the telltale screen.

- Stress the sand pack at 700–1,000 psi, or as necessary.

- Shut down and wait for one hour for the gravel-packing sand to set (with or without holding any pressure on the annulus).

- Re-stress the gravel-packing sand around the screen and liner at 700 psi, if necessary.

- If the stress pressure is satisfactory, prepare to get off the screen and liner carefully.

- Pick up the tubing string and pull 800–1,000 psi tension over the tubing weight.

- Right-hand rotate the tubing string with a pipe wrench at the rig floor for eight to twelve rounds at the tool to back off the right-hand releasing tool on the top of polished nipple. (You may see the pipe jump free.)

- Pick up the work string ten inches off the polished nipple and reverse circulate the wellbore clean two pipe volumes. (Take caution not to break the small wash pipe inside the screen and liner when picking up on the work string!)

- Start pulling the work string tubing slowly to avoid breaking the small wash pipe inside the screen and liner.

- Continue pulling the pipe at a slow speed to get the work string and the bottom-hole assembly out of the wellbore safely. (Slow down when pulling the small wash pipe; we are not in racing business!)

- Back off and lay down the small wash pipes in singles using pipe clamps; lay down the right-hand release sub and the piston assembly. (Hope that you did not leave anything in the screen and liner.)

- Note: it has happened that the small pipe threads flared or galled during the makeup, backing off fluid wash, which caused the loosening of the wash-down pipe inside the screen and liner.

- Any lost objects must be fished out. During one incident, the small pipe broke while coming out of the screen and liner and left three joints of one-inch IJ (integral joint) wash pipe in the well (Thompson field). The impression block indicated that the broken one-inch pipe landed outside the screen and liner in the annulus area. Two joints of one-inch pipe were fished out, and the third joint was left outside the telltale joint in the annulus.

- Pull the work string and wash pipe slowly out of the casing string to avoid losing tools.

- Stand back the work string, and lay down the wash pipe in singles using pipe clamps.

- Trip in the hole with an isolation packer (compression-type L-packer with lead seal assembly or O-ring type overshot).

- Tag and slack off to shear and to set the isolation packer over the polished nipple.

- The advantage of an isolation packer is as follows:

  a. It is easy to go through and check inside the screen and liner.

  b. It prevents gravel-packing sand from pushing back into the screen from the annulus because of gas surge.

- Make sure the isolation packer is set properly and packed without any gaps between the polished nipple and the isolation overshot.

- Pull out of the hole, and lay down retrieving tools.

- Run the production equipment, and put the well on production.

- Lift the production fluid slowly for the first five days to avoid fine formation into the gravel-packing sand.

****

The Concept of Conventional Gravel Packing in a Cased-Hole

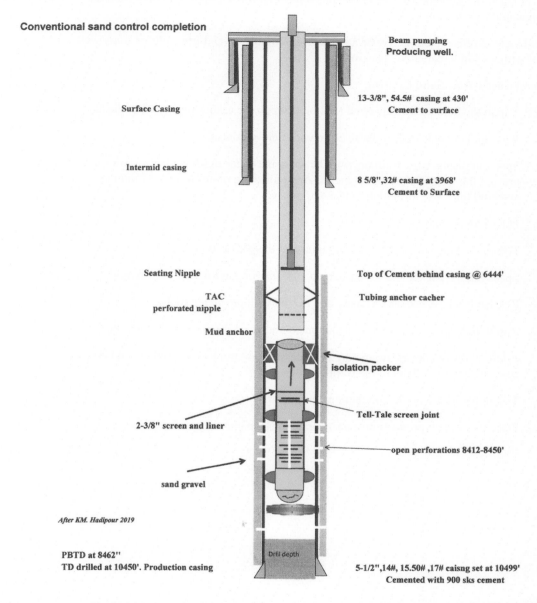

Conventional sand control completion

Beam pumping
Producing well.

Surface Casing

13-3/8", 54.5# casing at 430'
Cement to surface

Intermid casing

8 5/8",32# casing at 3968'
Cement to Surface

Seating Nipple

Top of Cement behind casing @ 6444'

TAC
perforated nipple

Tubing anchor cacher

Mud anchor

isolation packer

Tell-Tale screen joint

2-3/8" screen and liner

open perforations 8412-8450'

sand gravel

After KM. Hadipour 2019

Drill depth

PBTD at 8462"
TD drilled at 10450'. Production casing

5-1/2",14#, 15.50# ,17# caisng set at 10499'
Cemented with 900 sks cement

The conventional concept of sand-control method is to run and set the screen and liner across the gun-perforated casing and/or open hole without pumping gravel-packing sand in the well. This concept is based on using the coarse sand grains that flow out of a reservoir formation, which, hopefully, may pack off around the screen section.

The conventional gravel-packing method may apply to the wells with coarse reservoir sand grains based on the core samples or the wellbore sand samples.

Large formation-sand grains may be sufficient and good enough in order to run and set a sand screen without gravel packing. The coarse reservoir formation sand maybe used to protect the screen and liner and to avoid finer formation sand entering into the screen and liner without additional gravel packing.

This method of gravel packing was done in early oil-field gravel packing.

This method is done using the natural coarse formation sand that is produced by the reservoir formation to become packed off around the screen and liner without additional gravel packing. Installing the wire-wrapped screen without gravel packing in a cased-hole wellbore is impossible to justify. The formation sand usually contains fine and coarse gravel and may plug off the screen section (taking chances).

The success of conventional gravel packing in a cased-hole completion may not be too favorable because of early screen-section failure (plugged-off screen). The screen becomes plugged off with contaminated mud and fine solids and may cause large decline in the productivity.

The conventional method of completion may be applied in cased-hole oil-and-gas wells or water wells with low reservoir pressure, with or without additional gravel packing.

Conventional Cased-Hole Sand Control

a.   Trip in the hole with a full-sized rock bit through the casing string and in the hole.

b.   Wash and circulate the wellbore clean with field-produced water.

c.   Spot and displace the wellbore with heavy brine water to obtain hydrostatic fluid to prevent the formation from caving in. An application of gelled/brine fluid may keep and hold the loose formation in place, if needed.

d.   POOH and lay down the rock bit.

e.   Trip in the hole with the screen and liner and the packer-type assembly.

f.   Space out the packer and set the screen and liner in place across perforations.

g.   Release from the packer assembly and POOH with retrieving tools.

h.   Run in the hole with production equipment and place the well on production.

i.   Additional wellbore cleaning may be necessary to circulate the gelled fluid and any fine formation sand out of the wellbore.

j.   Trip in the hole with the production equipment.

k.   Place the well on production at a slow rate to clean up.

****

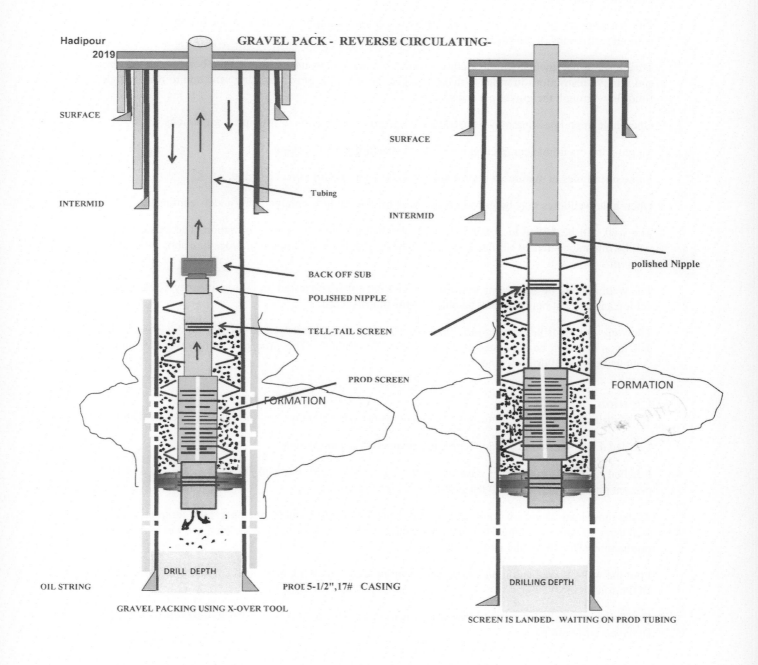

GRAVEL PACK - REVERSE CIRCULATING-

Hadipour
2019

SURFACE

INTERMID

Tubing

BACK OFF SUB

POLISHED NIPPLE

TELL-TAIL SCREEN

PROD SCREEN

FORMATION

OIL STRING

DRILL DEPTH

PROL 5-1/2",17# CASING

GRAVEL PACKING USING X-OVER TOOL

SURFACE

INTERMID

polished Nipple

FORMATION

DRILLING DEPTH

SCREEN IS LANDED- WAITING ON PROD TUBING

I.   Concept of Open-Hole Gravel-Packing Method

Open-hole gravel packing is an excellent technique and opportunity to successfully complete a well and to avoid unconsolidated formation sand from entering the wellbore.

Open-hole gravel packing is done in freshwater wells, oil wells, saltwater wells, or disposal wells.

The purpose of open-hole gravel packing is to obtain maximum production for higher profitability without restrictions of the casing holes.

This method may include window cutting, hole enlargement, or under-reaming the borehole to enlarge and remove restrictions in order to make a larger space for the gravel-packing sand and to permit high fluid flow through the gravel-packed screen.

Open-hole gravel packing may be used in a single-zone completion or multiple completions.

Single-zone open-hole completion is similar to standard gravel packing.

Some open holes in mature wellbores are stable for a successful gravel packing.

Open-hole wellbores may be gravel packed with or without hole enlargement (under-reaming).

The well will be drilled to a planned casing depth first above the target formation. The casing string will be run and cemented similar to normal standard drilling and completion procedure above the target reservoir.

The wellbore will be cleaned up to the top of the float collar and tested for casing integrity. Gamma ray/ ccl log and cement bond log will be obtained for future record.

The well will be drilled through the float collar and float shoe later with a rock bit to the target zone. The open hole will be circulated clean and logged for records. (This procedure will prevent formation damage because of drilling mud invasion.)

Caution

When drilling down through new pressurized reservoir, the well may flow and cause well-control issues.

I drilled a well outside of Bryan, Texas, for the open-hole completion.

I drilled the well to the target zone using 10 ppg drilling fluid; the well started flowing gas and 100 percent pipeline oil with 2,860 psi shut-in drill pipe pressure!

At the time, I was using a workover rig to drill and complete the well for the open-hole completion. The well came in while reaching TD (total drilling depth). I was told not to kill the well and put the well on production. (Hopefully, the well will die soon).

I put the well on production on 16/64″ choke with the 3⅞″ rock bit, eight drill collars, work string, and BOPs still on the wellhead.

One and a half years later, the well was still flowing and slowed down to thirty-five barrels of oil. (The BOP was still holding.)

****

Prepare to Run and Use a Hole Opener

The formation hole opener will be spaced and run to below the float show, and a pilot hole may be drilled to facilitate an under-reaming operation.

Sand-control techniques are carried out in the open holes using several methods.

Safety First

- Move in a workover rig and equipment to conduct open-hole sand control.

- Review the tubing string size, grade, and condition of the work string rating.

- Verify derrick rating and drill-line condition before rigging up on the well.

- Move in a workover rig to a dry and harden well location (if land work).

- Check the well location for hazards and overhead power lines. Move in and spot the fluid pump and a clean rig tank.

- Check and remove any fencing around the well prior to backing up the well.

- Use lock out and tag out, if necessary. Install windsocks!

- Check the location to ensure the ground is dry and hard (suitable to rig up). Use steel mats to support the rig from sinking into the soft sand and clay.

- All the rig crew must wear their personal protective equipment (PPE), such as hard-toed boots, hard hats, eye-safety glasses, working gloves, clean and dry long-sleeved shirt, and pants.

- Check for toxic gas. Never assume a well or location to be H_2S or CO_2 gas-free!

- Start move in and back up to the well. Use two spotters to guide the rig back to the wellhead. Do not run into the wellhead equipment and flow lines.

- Rig up on the well. Hang blocks/pipe elevators straight at the center of the wellbore.

- Do not rig up the rig within twenty feet of any power line.

- Tension test guy-wire anchors 23,000 lbs. If using steel-base beams, the beam anchors must be certified to stay in compliance (use API guidelines).

- No smoking within 150 feet from the wellhead.

- String out guy-lines to the safety anchors away from any power lines.

- Review the wellbore with a company man before you start working on a well.

- Check and calibrate the weight indicator to avoid accidents.

- Check the drill line and change if necessary.

- Use a body harness when working five feet above the rig floor.

- Get ready to work on the well (conduct safety meeting).

Wellbore preparation before under-ream drilling

Note: Under-reaming may have adverse results in some formations. Under-reaming may cause some reservoir formation to collapse and complicate the wellbore completion.

- Nipple up and test BOPs at 200 psi low and 5,000 psi high. (No shortcuts on BOP testing.)

- Never take shortcuts on safety. (It will catch up with you.)

- Keep the safety valve on the rig floor in full opening position with the wrench.
  Teach your people how to use safety valves during flow emergency.

  The TIW safety valve is designed to hold pressure from the bottom only.
  Finish rigging up the rig floor and handling tools.

  ► Prepare the tools and equipment to conduct under-ream drilling.

  ► Under-reaming requires a lot of planning, experience, and very close supervision. The operation may be a complicated and lengthy process, which requires good equipment and experienced operators.

Under-reaming is done in two methods:

a. Under-reaming through a pilot hole
   Under-reaming through a pilot hole is safer and less complicated.

b. Under-reaming full formation interval
   Under-reaming solid rock formation is time-consuming and more expensive.

Open-Hole Gravel Packing Using an Under-reamer

The open-hole gravel-packing method using an under-reamer is an excellent concept of well completion. The under-reamed reservoir formation will be fully opened without steel casing restrictions. Maximum production can be achieved from this open-hole completion technique.

Open-hole gravel packing is primarily used in freshwater wells, saltwater disposal, as well as oil well gravel packing.

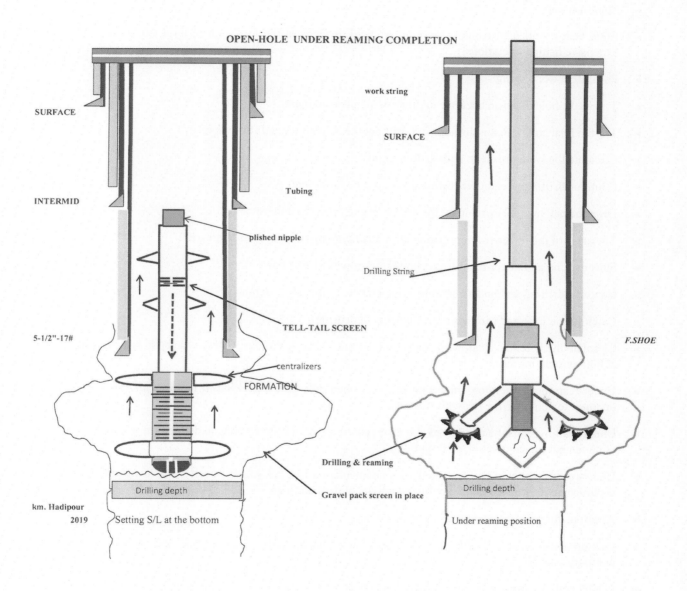

OPEN-HOLE UNDER REAMING COMPLETION

Preparing for Under-reaming Operation—Short Procedure

Trip in the hole to enlarge a borehole. (Open pilot hole from 6-⅛″ to 16″ of the hole.)

► Full-sized rock bit and bit sub (preferably a blade-type under-reamer rather than cone)

► 16″ calipered under-reamer

► Stabilizer sub

► Eight heavy wall drill collars (Drill collars tend to break at the tool joint in large holes and cause expensive fishing operations!)

► Ph-6, P110 work string

► Trip in the hole; keep an accurate drilling string measurement.

► Tag the float collar. Ream through and drill out the float collar and float shoe.

► Circulate the wellbore clean with drilling fluid.

► Continue drilling to eight feet below the float shoe (new hole).

► Drop the ball and pressure up on the drill string to open the under-reamer.

► The arms of the under-reamer open or close hydraulically with hydraulic pressure.

► Start drilling and under-ream from eight feet below the float shoe.

► Continue drilling and under-reaming the open hole to the target depth.

► Catch formation-sand samples every five feet of drilling.

► Circulate the borehole clean with heavy gelled fluid or heavy viscous fluid to efficiently remove solids at high circulating rate.

► Under-reaming is the most important phase of open-hole gravel packing. It requires an experienced operator to complete the job with success.

► Under-reaming must be conducted at a slow rate to obtain uniform borehole diameter through the use of nondamaging completion drilling fluid.

► Under-reaming should be conducted with nondamaging low gel, high viscosity, and low fluid loss to avoid formation damage and maximize removal of drill cuttings.

► After reaching the target depth, circulate the borehole clean.

► Pick up and drop the ball to close up the reamer. (Under-reamer will close and open with hydraulic pump pressure.)

► POOH with the work string and the bottom hole assembly. Lay down the under-reamer.

► Caliper the borehole after under-reaming to determine the actual borehole and to ensure there is no formation sloughing.

► Trip in the hole with a mule shoe joint on the drill string. Wash and reverse circulate the borehole clean at high circulating rate using heavy gelled fluid.

► Pull the drill string in clear three hundred feet. Shut down and check the well for flow.

► Run in the hole with caliper log to verify the quality of the under-reaming operation.

► Obtain accurate borehole measurements for the sand gravel.

► Reverse circulate the borehole clean and spot with fresh, heavy, viscous gelled fluid to keep the wall from caving in and to increase hydrostatic head on the borehole.

► Pick up and wait for one hour and tag the bottom, if necessary.

► If the wellbore is stable, you may run the screen and liner tool and gravel pack in place.

► POOH with the work string while keeping the hole full with heavy fluid. Lay down the tools.

► Space out the screen and liner assembly from the bottom of the open hole up to sixty feet into the casing string.

Trip in the hole with the screen and liner assembly:

— Wash down shoe (reverse circulating shoe)
— The sand screen section with bow spring–type centralizers, spaced forty feet apart (for the open-hole section)
— Blank pipes with centralizer above the screen to fit inside the casing string
— Telltale screen joint
— Blank pipes with weld-on centralizers to fit casing
— Polished nipple
— Back-off sub
— Work string

► Trip in the hole and set the screen across the open-hole interval using bow spring–type centralizers.

► Make sure sufficient gravel-packing sand is available to do the job.

► Rig up the gravel-packing unit.

► Install lines and check the flow direction to avoid pumping sand the wrong way.

► Mix and displace gravel-packing sand around the screen and liner from the bottom of open hole to the telltale screen.

► Pump, pack, and stress larger-sized gravel-packing sand in the open hole and around the screen and liner assembly, similar to cased-hole gravel packing.

► Do stress packing as required.

► Release from the screen and liner and reverse out hole clean.

► POOH and lay down excess tools.

► Trip in the hole with an isolation packer using internal lead seal or O-ring type overshot.

► Wash the wellbore, if necessary, using filtered fluid.

► Place the well on production at a slow rate.

****

Casing Open-Hole Method

Open-hole gravel packing can be done in cased-hole wellbores through a section hole in the casing string.
A large section of casing will be milled out and removed. The open-hole gravel packing will be conducted similar to open-hole drilling operations.

Read the completion and drilling techniques published by the author for reference.

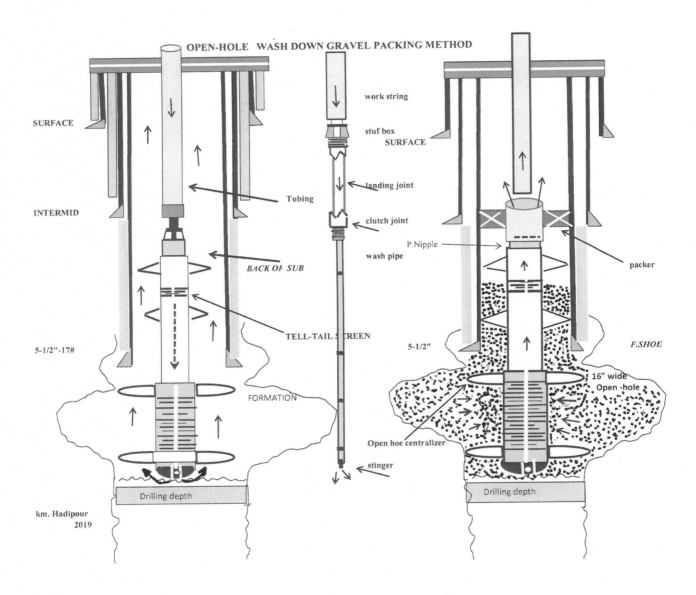

OPEN-HOLE WASH DOWN GRAVEL PACKING METHOD

II.   Concept of Slurry Gravel-Packing Method

The concept and application of slurry gravel packing may be different from water-pack gravel packing method!

Slurry gravel packing is often referred as gelled-slurry gravel-packing method.

The carrying fluid is made of high-concentration viscous gel polymer. The highly viscous fluids used in slurry packing have a tremendous sand-carrying and suspension ability.

The thought process behind highly viscous fluids or gelled-sand slurry is to carry, squeeze, and deposit unbelievably high sand concentrations of six pounds to fifteen pounds of gravel in one gallon of high-viscosity carrier polymer fluid in order to achieve maximum sand pack and minimize fluid loss to the formation. (I have mixed feelings about the whole concept.)

This thought process unfortunately may not always be successful because of gravel-packing design, sand-out problems, and mechanical failures during the slurry gravel packing through the gun-perforated casing string. If the job is done successfully, the volume and rate is very low (not a tight gravel packing). If the job fails because of sand-out problems, the sanded wellbore will be left without gravel-packing sand.

A typical slurry gravel-packing work is outlined below:

1. Move in and rig up the workover rig (a pulling unit).

2. Spot the rig pump and tank. Steam clean the rig tank and fill up with clean water.

3. Kill the well with filtered brine, if necessary. If the well goes on vacuum and is unable to circulate the wellbore, you may prepare to isolate the perforations before running the scrapers. Scraping loose the contaminated elements is very bad for gravel packing!

4. POOH with the production equipment in the well to gravel pack.

5. Trip in the hole with a rock bit and a set of casing scrapers in tandem. Tally (strap and test) the tubing string at 5,000 psi going in the hole. Run and scrape the casing string from the surface to the top of the open perforations. Circulate the wellbore clean using filtered fluid. Check and analyze the elements that are circulated up the hole.

6. Acid pickle the tubing and casing string to remove formation deposits, if necessary.

7. POOH and stand back tubing string. (If the tubing string contains paraffin, you may consider steam cleaning the tubing above the ground to remove the entire paraffin content.
See the tubing cleaning procedure published by the author.)

8. Rig up the wireline unit to run and set a bridge plug. Nipple up and test the casing lubricator.
RIH with a gauge ring and junk basket, and clear and drift the casing to twenty feet below the open perforations. POOH and lay down the wireline tools.

9. GIH with a cast-iron bridge plug. Correlate and set the bridge plug twelve feet below the open perforations, or as recommended. (The cast-iron bridge plug is the best footing source to land the screen and liner.)

10. POOH and rig down the wireline tools and equipment.

11. Trip in the hole with a bit and scraper to clean up the wellbore. Tag the bridge plug and correlate the pipe tally with the bridge-plug depth. Pick up and circulate the borehole clean with kill fluid.

12. POOH with the tubing string while keeping the hole full of fluid.

13. Prepare the gravel-packing tools and equipment to run the screen and liner.

14. Keep an accurate pipe tally of anything that goes into the well.

15. Rabbit and drift anything that goes into the well.

16. Inspect and trip in the hole with the following gravel-packing assembly:

    a.   Blind bull plug.

    b.   Screen section/sections (internal and external plastic coated with three feet overlap screen above and three feet overlap screen below).

    c.   Blank pipe, as required (external and internal coated with centralizers and spaced twelve to fifteen feet apart). Make up screen and gravel-packing assembly by hand and friction wrenches to avoid damages.

    d.   Gravel pack hook-up nipple (GHUN) with the releasing tool.

    e.   Ten feet of safety joint (short pup joint).

    f.   R-type isolation gravel-packing packer (with new and functional bypass elements).

    g.   Test, pickle, and clean the work string tubing (spaced out to three feet above the rig floor). Always make the tubing string by hand first, then apply hydraulic tubing tongs with proper torques.

    h.   Apply light-thread dope (lubricant) to pin ends only.

****

17. Trip in the hole with the screen and liner and gravel-packing packer assembly slowly to avoid fluid surging. Stop above the cast-iron bridge plug (CIBP), check the string weight and tubing tally, and slack off slowly.

18. Tag up on the bridge plug and check the pipe tally (correlate the tubing string and BHA to the bridge-plug depth).

19. Pick up to the packer stroke length and right-hand motion. Rotate to set the packer using a pipe wrench only. (Rotate the pipe at the surface enough to get one-fourth-inch turn at the tools to set the packer.) Slack off on the packer and set down to 9,000 lbs. to pack-off isolation elements.
The slack-off weight on the packer depends on the size of the casing string.

20. Fill up the annulus and test packer at 600 psi, or as required. (Keep the tubing open when testing the annulus.)

21. Rig up the gravel-packing equipment (sand pots, filter, and slurry gravel-packing unit and equipment). The slurry gravel-packing equipment consists of a large trailer-mounted sand/polymer blender or skid-mounted equipment, and it may require a larger space compared to a little water-pack gravel-packing unit. In the offshore operation, a large boat, like a ship, is used to conduct sand slurry gravel-packing operations.

22. Lay down and make up the steel flow lines. Flush and test the lines at 5,000 psi, or as required. (The rig crew needs to stay away from the lines while it is under pressure.)

23. Pressure up on the annulus at 600 psi. Pick up on the work string and open bypass on the isolation packer slowly.

24. Establish circulation at a rate of two barrels per minute with clean, filtered completion fluid. Shut down and allow the pressure to drop to zero. Slack off and close the bypass. (Never surge the pressure.)

25. Establish the injection rate and pressure with completion fluid. Record the rates and injection pressure through the gun perforations to evaluate wellbore-treatment rates and pressures.

26. Check that the bypass is closed on the packer and hold at 600 psi on the annulus. Shift to squeeze mode by slacking off 10,000 lbs. on the packer, or as required. Prepare to slurry pack by pumping sand and polymer slurry down the tubing string and out into the gun-perforated holes and around the screen and liner. (I prefer batch mixing to obtain a consistent sand slurry.)

27. Continue pumping using gelled fluid (completion fluid) at two barrels per minute, or as required, followed by sand polymer slurry volume as needed to slurry the reservoir formation. If the sand-slurry squeeze is not obtainable because of unconsolidated reservoir formation, you may have to carefully cut back on the volume, rate, and gel concentration (like slick water) in order to achieve successful tight gravel packing.

28. As soon as the sand slurry pack is completed, switch to slurry displacement using completion slick fluid. (Never over-displace or under-displace the slurry gravel packing.)

29. Displace and pack the sand slurry to 1,500 psi and re-stress the sand pack to ensure proper sand out.

30. Wait on the sand pack to heal for fifteen minutes, if necessary. Pressure up and hold to 600 psi on the annulus and packer assembly.

31. Pick up on the bypass with 600 psi back pressure and reverse circulate the excess sand slurry, if any. Circulate the borehole clean with two tubing volumes.

32. Pick up to engage the clutch on the gravel pack hook-up nipple (HUN).

33. Rotate eight to ten right-hand rotations to release from the hook-up nipple–releasing tool.

34. Circulate the wellbore clean with two volumes, if necessary.

35. Trip out the hole with a bottom-hole assembly (BHA) slowly. Lay down the packer and the bottom-hole assembly. (Pull the packer slowly to avoid swabbing sand into the hole.)

****

Trip in hole with production equipment after gravel packing as follows:

1. Trip in the hole with a correct-sized overshot for the hook-up nipple.

2. A landing nipple

3. The production isolation packer

4. Landing nipple profile

5. Test and run the production tubing string to the surface, if necessary.

6. Finish going in the hole with the production packer and the overshot assembly. Slack off with the overshot and engage hook-up nipple overshot. Tag and set down to 6,000 to 9,000 lbs. to ensure the hook-up nipple is fully engaged.

7. Space out the tubing string and land tubing in the hanger assembly.

8. Rig up the wireline equipment. RIH with a wireline plug to land in the nipple below the packer. Pressure up on the tubing string to 2,000 psi to set the packer. Hold and check the pressure for fifteen minutes.

9. POOH and lay down the wireline plug and tools.

10. Test the packer, if necessary, and put the well on production.

11. Flow the well back after twenty-four hours at a slow rate.

****

Comments about High-Viscosity Polymer Fluid in Gravel Packing

The idea of high-pressure hydraulic pumping rate is no different from fracturing stimulation. The higher the rate, the higher applied hydraulic pumping pressure is.

A high concentration of six to fifteen pounds per gallon or more gravel-packing sand in one gallon of gelled polymer fluid is not a gravel pack at all and should be called a frac-pack stimulation method.

A high-concentration pack requires higher hydraulic pressure to force the gravel-packing sand through the gun-perforated casing holes and into the formation. Sweeping and rearranging the formation sand behind the perforated tunnels is a major concern of permeability damage.

The supporters of high-viscosity slurry gravel packing believe that high sand concentration pack will prevent formation damage and avoid gravel intermix with the reservoir formation.

High sand concentration slurry gravel packing may prevent pumping large volumes of fluid into the productive reservoir.

The slurry gravel-packing sand is an ideal method of breaking down and gravel packing tight reservoir formation sand but not loosen unconsolidated formation and partially depleted wells.

The fracturing record proves that tighter formations can highly be productive of oil-and-gas if it is fractured-packed.

Frac-pack sand-control techniques have been used in several unconsolidated formation zones in the Gulf Coast area with mixed results (waste of money and damaging formation).

It is unfortunate that no formal study is done on the effects of high-viscosity, high-gravel concentrations, and high-rate slurry packs on unconsolidated reservoir formations. (The results of such study on the subject may or may not be too impressive.)

****

The only study done by the author while gravel packing and fishing several slurry-packed screens and liners in the offshore and land wells concludes the following:

- Slurry gravel packing appears to be an effective sand pack in semi-tight reservoir formations.

- Frac-packing the tight reservoir formation proved to be superior to water-based sand gravel-packing methods because of viscous-fluid-carrying capabilities and hydraulic-pressure breakdown of the formation.

- Viscous, slick fluid is capable of holding and carrying higher sand concentration superior to water-packing methods.

- It is also found that the carrier gel polymer (in gel slurry packing) is still in the form of mid-hard honeycomb substance after several years in the shallow- and low-temperature wells.

- Several fished screens and liners out of wellbores with slurry packing indicate that small to no gravel-packing sand is left above or across the screens and liners.

- Pulled screens and liners out of the wellbores were found with fine formation deposits and scale in the form of cementation, which suggests the lack of gravel-packing sand (in gelled slurry) to protect the screen and liner.

- Dehydrated mud and sand grains were packed together on the liner above the screen.

- The shallow, unconsolidated wells of the Gulf Coast area with low temperature and low pressure were not suitable for heavy gelled slurry gravel packing.

  The slurry-packing method may be more effective in deeper, tight reservoir formation wells with lower porosity and permeability.

- The fracturing pressure in unconsolidated reservoirs is low for slurry packing.
  The high rate, high-hydraulic pressure, and high viscosity of the carrying gel fluid with high sand concentration will either cause sand-out or may cause deep formation-sand sweep with large cavities into the reservoir rock, causing formation-sand rearrangement.

- Often it would take several days and repeated sand batches to achieve a slurry gravel-packing objective in low-pressure, unconsolidated reservoir formations.

  After two days of gravel packing, the pumping rate had to be cut to one barrel per minute. Heavy viscous gel had to be cut back to near slick water to obtain gravel-packing results and in order to get off the well (Goose Creek Bay, Hankamer Field, and offshore south Louisiana).

- Chemical bridging material is often used with gel slurry packs, which is not completely soluble and not compatible with reservoir rock and the well-stimulation process.

- It is impossible to reach the bridging material with breaking fluid without washing, sweeping, or pushing the gravel-packing sand away from the wellbore.

  Forty-five hours after slurry packing, the well was flowing back with sand and unbroken gel fluid.

- The biopolymers or guar gums and HEC pills are used along with viscous gravel-packing fluid as bridging materials in order to make the compound coagulate or thicken to semisolid state.

- Bridging materials are used in completion fluid to obtain more effective fluid-loss control.

- A mixing blender is mounted on a flat trailer and used in high-viscosity gravel-packing squeeze method.

- The gravel-packing sand is mixed in a large blender that is equipped with sharp-blade paddle wheels to be mixed without any crushing controls. The mixed slurry will then be fed to a high-pressure hydraulic plunger pump to be forced down the hole without any concern about the physical condition of the gravel-packing sand on downstream.
  (No doubt that the gravel is subject to a high degree of crushing smaller-than-the-screen slots!)

- Some of the additives are mixed and pumped on fly and did not shear at all.

- These semisolid compounds are difficult to break and may cause reduction in reservoir productivity after a few months.

- In addition, the breaker agents, such as HCL or HF acids, will be added to the compound in order to dissolve semisolid polymer materials.

- The HCL and HF acid will react and dissolve with gravel contaminants and will break down the sand grains to smaller particles, and they are actually unsafe to use. (Never use acid if you are using glass beads.)

- Sand crushing—the sand-transporting method, mixing of sand in tubs and blenders with high-powered hydraulic rates cannot be avoided by blender operator at all.

- Crosslinked gel fluids, such as those used for gravel packing, may provide little fluid loss to the formation but will also prevent tight gravel packing in the formation.

- Shrinkage and voids may occur after high-viscosity polymer breaks down, which can be up to 35 percent of the total volume of the slurry pack. You will leave large voids within the sand pack and loosen gravel around the screen and liner assembly. Premature sand bridging may occur that will leave unpacked sand around the screen. (This is a leading trouble for the future.)

- It is believed that the thick high-rate gel-slurry gravel pack will tend to intermix with formation sand and may cause skin damages to formation and gravel packing; it may cause a decline in fluid productivity gradually with time.

- Honeycomb residue of thick crossed linked carrier gel fluid in low bottom-hole temperature wells may result in blockage behind the casing and reduce the flow rate and fluid productivity.

- Sand-grain crushing and intermixing is possible during high-rate sand concentration of eight to fifteen pounds per gallon of sand in one gallon of crosslinked viscous fluid.

- Sand bridging of slurry gravel packing is not preventable; it is another major problem during slurry pumping.

- During a sand screen, the operator will be forced to shut down without proper fluid flushing. This may occur as a result of higher pumping rate, high pump pressure, and heavy sand concentration through gun-perforated casing holes.

- After the sand screen, the gel slurry may break down with time, and the wellbore may go on vacuum, causing the carrier fluid and sand slurry to disappear into the created cavities of the formation, leaving the screen and liner assembly practically without sand protection.

- Slurry gravel packing yields a good production after gravel packing and will decline fluid rapidly later because of mixed formation sand with gravel-packing sand.

- In successful gravel packing, the perforation tunnels and the formation behind the casing string must be packed with clean resieved gravel sand without fracturing the formation.

- The annulus behind the screen and liner assembly must be fully packed and protected with sand to prevent fine solids from passing through the screen section across the open perforations.

- It is difficult to verify a dependable quality and quantity of sand pack around the screen and liner in the slurry-packing process.

****

Through-Tubing Gravel Packing

Through-tubing sand-control gravel packing is an innovative sand-control solution in some oil-and-gas wellbores with unconsolidated formation sand and downhole mechanical problems.

Through-tubing gravel packing is normally carried out in the older wellbores with formation-sand problems, where it is too expensive to move a workover rig, pull the production equipment out of the wellbore, and gravel pack the well properly.

Through-tubing installation is based on a small-diameter wire-wrapped screen ranging from ¾″ to 1-½″ outside diameter. The concept of through-tubing is to run and place smaller wire-wrapped screen inside the existing production string without having to pull the production equipment out of the well because of high-remedial workover cost.

This method is applied in various oil- and gas-well completions and interventions offshore as well as onshore using electric wireline, slickline, or coiled-tubing units.

Some through-tubing installation may actually last long enough considering the sizes and types of screen and liner. Through-tubing sand pack may yield good production volumes if it is done properly.

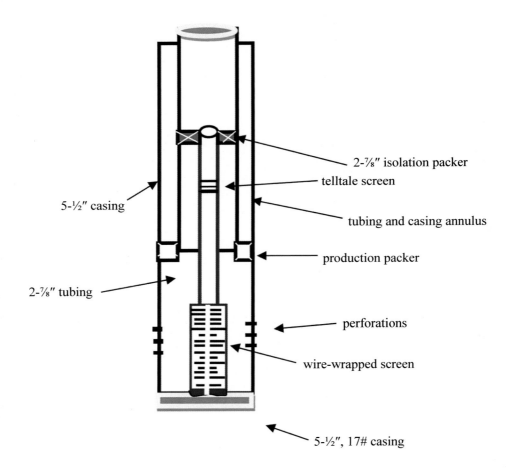

Note:
You may carefully evaluate the wellbore condition and the cost before considering a through-tubing gravel-packing project (future cost of workover and fishing work).

Through-tubing gravel packing may be viewed better than patching or putting a band-aid on a well for repair.

Through-tubing gravel packing is another innovative method of remedial work to repair or extend the useful life of an oil-and-gas production further.

Tools and equipment used in the through-tubing gravel packing are normally small in size, lightweight, and subject to become damaged, twisted, or parted, and may cause expensive fishing operations.

The tools can be run or pulled out of a well using a coiled tubing unit and an electric wireline. (Expect higher fishing and operating cost than a normal gravel packing.)

There are various methods of through-tubing gravel packing:

—wash-down method
—reverse circulating method

- The wash-down method uses small-sized work string or coiled tubing.

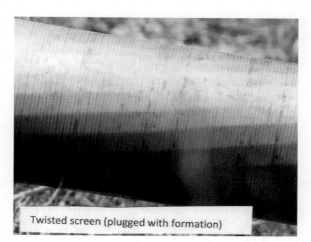

Twisted screen (plugged with formation)

1. Trip in the hole to clean up the wellbore to plug back depth using a coiled tubing unit. Coiled tubing is an excellent choice for wellbore cleaning, gravel packing, and well control. Rig up to wash and circulate the wellbore clean with heavy filtered brine.

2. Obtain formation-sand samples for analysis, if necessary.

3. Establish pump-in rate through the open perforation tunnels.

4. Pump and Braden-squeeze selected resieved sand gravel through the open perforation at low rate and low pressure (similar to prepack gravel packing).

5. Slack off and reverse clean at the plug-back depth.

6. Pull up the hole and shut down; wait for one hour or longer to verify the borehole condition.

7. If the wellbore stays clean and stable, prepare to run and set the screen and liner.

8. POOH with the coiled tubing while keeping the hole full.

9. Trip in the hole with the screen and liner. Tag the bottom and release from BHA.

10. POOH with the coiled tubing or electric wireline running tools.

11. Rig up the gravel-packing equipment and install the pumping line to the tubing string.

12. Calculate the sand volume. Pump and displace the gravel down the tubing string and around the screen and liner as required.

13. Check the screen and liner, if necessary, to make sure it is clear and clean to bottom.

14. Put the well on production.

15. If the wellbore is unstable, continue to step 16.

16. Pump and displace clean gravel-packing sand from the plug-back depth to thirty feet above the open perforation.

17. POOH with the pipe while assembling the screen and liner joints.

18. Make up the screen and liner, and trip in the hole as mentioned before.

19. Tag the sand and wash down the screen and liner in place.

20. Release from the screen and liner. POOH and rig down the coiled tubing or work string.

21. Put the well on production.

****

Reverse Circulating Method

Through-Tubing Gravel-Packing Procedure (Through-Tubing)

a. Inspect and remove all the hazards found in the well's location.

b. Rig up the coiled tubing; check and test the coiled tubing tools and equipment. Read and record shut-in tubing and casing pressure.

c. Check and test all tools and equipment.

d. Trip in the hole with a wash-tip on the coiled tubing. Wash down and clean up going into the wellbore. Wash and circulate the sand and solids to plug-back depth below the open perforations.

e. Circulate the wellbore clean to plug-back depth below the open perforations (footing).

f. Tag the bottom and mark the pipe for reference.

g. Pull up the coiled pipe to four hundred feet above the reference point.

h. Shut down and wait for one hour or longer to find out if the wellbore is stable.

i. If the hole is clean and stable, continue POOH with the coiled tubing while keeping the hole full. Pull the pipe slowly to avoid swabbing.

j. Pick up and trip in the hole with the screen and liner assembly on the coiled tubing or electric wireline at a moderate rate. (Make certain that the screen section is long enough to overlap and cover the entire perforation.)

k. Finish going in the hole with the screen and liner assembly slowly while going through the production tubing in the well.

l. Tag the bottom and check to make sure the landing point is correct.

m. Set the screen and liner at the bottom.

n. Release from the screen and liner assembly.

o. POOH with the coiled tubing string while preparing to pump through the tubing string.

p. Install and hook up the pumping line to the production tubing (2-⅜″, 2-⅞″).

q. Establish pumping rate, as required. Rate of one or two barrels per minute (BPM).

r. Calculate required sand volume to cover the screen and liner and void above. Mix, pump, and displace gravel pack around the screen and liner, pumping either gelled gravel slurry or water-pack gravel packing.

s. Displace the sand slurry with filtered produced water.

t. Check the wellbore to make sure it is clear and clean.

u. Put the well on production. Flow the well with a small choke.

I hope you enjoyed reading and learning!

The Author would like to thank Mark Mitchell and Chase Mitchell of Mitchell Industries, Baytown, Texas for providing pictures used within this book.

Going in hole with screen(unwrapping plastic)

Full of sand

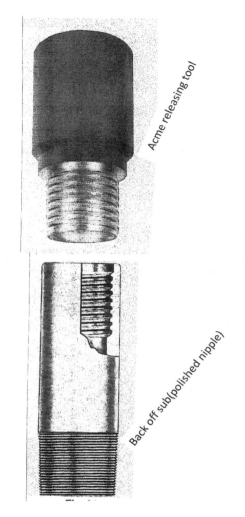

Acme releasing tool

Back off sub(polished nipple)

Unwrapped screen

Unwrapped screen

Flowing sand and solids from well

Printed in the United States
By Bookmasters